W9-ABO-144

LONDON MATHEMATICAL SOCIETY LECTURE NOTE SERIES

Editor: PROFESSOR G. C. SHEPHARD, University of East Anglia

This series publishes the records of lectures and seminars on advanced topics in mathematics held at universities throughout the world. For the most part, these are at postgraduate level either presenting new material or describing older material in a new way. Exceptionally, topics at the undergraduate level may be published if the treatment is sufficiently original.

Prospective authors should contact the editor in the first instance.

Already published in this series

1. General cohomology theory and K-theory, PETER HILTON.
2. Numerical ranges of operators on normed spaces and of elements of normed algebras, F. F. BONSALL and J. DUNCAN.
3. Convex polytopes and the upper bound conjecture, P. McMULLEN and G. C. SHEPHARD.
4. Algebraic topology: A student's guide, J. F. ADAMS.
5. Commutative algebra, J. T. KNIGHT.
6. Finite groups of automorphisms, NORMAN BIGGS.
7. Introduction to combinatory logic, J. R. HINDLEY, B. LERCHER and J. P. SELDIN.
8. Integration and harmonic analysis on compact groups, R. E. EDWARDS.
9. Elliptic functions and elliptic curves, PATRICK DU VAL.
10. Numerical ranges II, F. F. BONSALL and J. DUNCAN.
11. New developments in topology, G. SEGAL (ed.).
12. Proceedings of the Symposium in Complex Analysis, Canterbury 1973, J. CLUNIE and W. K. HAYMAN (eds.).
13. Combinatorics, Proceedings of the British Combinatorial Conference 1973, T. P. McDONOUGH and V. C. MAVRON (eds.).
14. Analytic theory of abelian varieties, H. P. F. SWINNERTON-DYER.
15. Introduction to topological groups, P. J. HIGGINS.
16. Topics in finite groups, TERENCE M. GAGEN.
17. Differentiable germs and catastrophes, THEODOR BRÖCKER and L. C. LANDER.
18. A geometric approach to homology theory, S. BUONCRISTIANO, C. P. ROURKE and B. J. SANDERSON.
19. Graph theory, P. J. CAMERON and J. H. VAN LINT.
20. Sheaf theory, B. R. TENNISON.

London Mathematical Society Lecture Note Series. 21

Automatic continuity of linear operators

ALLAN M. SINCLAIR

CAMBRIDGE UNIVERSITY PRESS
CAMBRIDGE
LONDON NEW YORK MELBOURNE

Published by the Syndics of the Cambridge University Press
The Pitt Building, Trumpington Street, Cambridge CB2 1RP
Bentley House, 200 Euston Road, London NW1 2DB
32 East 57th Street, New York, NY 10022, USA
296 Beaconsfield Parade, Middle Park, Melbourne 3206, Australia

Library of Congress Catalogue Card Number: 74-31804

ISBN: 0 521 20830 0

First published 1976

Printed in Great Britain
at the University Printing House, Cambridge
(Euan Phillips, University Printer)

Contents

Introduction

In these notes we are concerned with algebraic conditions on a linear operator from one Banach space into another that force the continuity of the linear operator. The main results are in the theory of Banach algebras, where the continuity of homomorphisms under suitable hypotheses is part of the standard theory (see Rickart [103], and Bonsall and Duncan [18]). The continuity of a multiplicative linear functional on a unital Banach algebra is the seed from which these results on the automatic continuity of homomorphisms grew, and is typical of the conditions on a linear operator that imply its continuity. Homomorphisms, derivations, and linear operators intertwining with a pair of continuous linear operators are the most important general classes of linear operators whose automatic continuity has been studied. These notes are an attempt to collect together and unify some of the results on the automatic continuity of homomorphisms and intertwining operators.

The most important results in these notes are in sections 4, 6, 8, 9, 10, and 12 of Chapters 2 and 3. The guiding problem behind Chapter 2 is to find necessary and sufficient conditions on a pair (T, R) of continuous linear operators on Banach spaces X, Y, respectively, so that each linear operator S from X into Y satisfying $ST = RS$ is continuous (Johnson [58]). The equivalent problem for homomorphisms is to find necessary and sufficient conditions on a pair of Banach algebras A and B so that each homomorphism from A into B (or onto B) is continuous (Rickart [103, §5]) (Chapter 3). Chapter 1 contains the general technical results on which Chapters 2 and 3 are built, and in Chapter 4 the continuity of positive linear functionals on a Banach $*$-algebra is discussed.

Throughout these notes all linear spaces will be over the complex field unless the space is explicitly stated to be over another field, and all linear operators will be complex linear except in Section 7 where we consider ring, i. e. rational linear, isomorphisms between semisimple Banach algebras. Many of the results do hold for real Banach spaces but we shall not consider them. Attention will be restricted to Banach spaces, and automatic continuity results for Fréchet spaces, and other topological linear spaces are not discussed. For example, the weak continuity of derivations on a von Neumann algebra (Kadison [70]), and the uniqueness of the Fréchet topology on the algebra $\mathbf{C}[[\chi]]$ of all formal power series in one indeterminate χ over the complex field $\mathbf{C}$ (Allan [1]) are omitted. Zorn's Lemma will be assumed throughout these notes and plays a crucial role in the existence of counter examples (Johnson [58, p. 88]). Axiomatic systems that imply the continuity of all linear operators between two infinite dimensional Banach spaces are beyond the scope of these notes (Wright [131]).

The reader will be assumed to know the basic theorems of functional analysis and elementary Banach algebra theory. In particular the following two results will be used frequently without reference: a finite dimensional normed linear space is complete; each linear operator from a finite dimensional normed linear space into a normed linear space is continuous. In certain sections deeper results from algebra and analysis are assumed. In Section 3 the properties of divisible, injective, and torsion modules over the principal ideal domain $\mathbf{C}[\chi]$ of all polynomials in an indeterminate χ over the complex field $\mathbf{C}$ are used in obtaining a discontinuous intertwining operator (Cartan and Eilenberg [22], Kaplansky [138]). The decomposition of a torsion module over $\mathbf{C}[\chi]$ is used in the proof of Theorem 4.1 (see Hartley and Hawkes [144]). The spectral theorem for normal operators on a Hilbert space is used in Section 5. The Wedderburn theory for finite dimensional semisimple algebras over $\mathbf{C}$ is required in Section 7 when we consider ring isomorphisms between semisimple Banach algebras (Jacobson [51]). In the same section, and in Section 8, we apply the single variable analytic functional calculus to an element in a Banach algebra (Bonsall and Duncan [18], or Rickart [103]). Elementary properties of field extensions, and the embedding of a domain in its field of fractions are used in Section 8 (Jacobson [50]). Section 12 requires elementary properties of C*-algebras (Dixmier [34]).

We shall now describe the contents of the various sections in more detail. The first section contains the basic properties of the separating space $\mathfrak{S}(S)$ of a linear operator S from a Banach space X into a Banach space Y, where

$$\mathfrak{S}(S) = \{y \in Y: \text{there is a sequence } (x_n) \text{ in } X \text{ with } x_n \to 0 \text{ and } Sx_n \to y\}.$$

The closed graph theorem implies that S is continuous if and only if $\mathfrak{S}(S) = \{0\}$. It is this equivalence that has made the separating space a useful technical device in automatic continuity problems. The most important result in this section is Lemma 1.6, which is used in Sections 4 and 11 (Johnson and Sinclair [69], Allan [1], Sinclair [118]). Lemma 1.6 shows that if (T_n) and (R_n) are sequences of continuous linear operators on Banach spaces X and Y, respectively, and if $ST_n = R_nS$ for all n, then there is an integer N such that

$$(R_1 \dots R_n \mathfrak{S}(S))^- = (R_1 \dots R_N \mathfrak{S}(S))^-$$

for all $n \geq N$.

In Section 2 we consider discontinuity points of an operator which leaves a large lattice of closed linear subspaces of a Banach space invariant. This idea seems to have been first used by Bade and Curtis [7], though not under this name, and has subsequently been exploited by many authors (for example Curtis [27], Gvozdková [48], Johnson [58], [63], [64], [66], Johnson and Sinclair [69], Ringrose [105], Sinclair [117], Stein [121], [122], Vrbová [128]). The conclusion when this method is used is that the discontinuity is concentrated in a subspace that is small in a technical sense associated with the lattice.

In Section 3 we prove that there exists a discontinuous linear operator S from a Banach space X into a Banach space Y satisfying $ST = RS$ under two additional hypotheses on T and R, where T and R are in $\mathcal{L}(X)$ and $\mathcal{L}(Y)$, respectively, and $\mathcal{L}(X)$ is the Banach algebra of continuous linear operators on the Banach space X (Johnson [58], Johnson and Sinclair [69], Sinclair [116]).

In Theorem 4.1 necessary and sufficient conditions are given on the pair (T, R), when R has countable spectrum, so that each linear operator S from X into Y satisfying ST = RS is continuous (Johnson and Sinclair [69], Sinclair [116]).

In Section 5 the operators R and T are assumed to be normal operators on Hilbert spaces and S intertwining with them is decomposed into continuous and highly discontinuous parts (Johnson [58]). The automatic continuity results that hold for other operators T and R with suitable spectral decompositions are not discussed in these notes (Johnson [58], Johnson and Sinclair [128], Vrbová [128]).

In Section 6 enough of the theory of irreducible modules over a Banach algebra is developed to prove the uniqueness of the complete norm topology of a semisimple Banach algebra (Corollary 6.13) (Johnson [59]). A full discussion of irreducible modules over a Banach algebra (and irreducible representations of a Banach algebra) may be found in Rickart [103] or Bonsall and Duncan [18]. Our proof of Theorem 6.9, on which the uniqueness of the complete norm topology of a semisimple Banach algebra depends, is no shorter than Johnson's original proof [59], but by basing it on Section 2 its relation to other automatic continuity proofs is emphasized. In Theorem 6.16 some properties of the spectrum of an element of the separating space of a homomorphism are given (Barnes [13]). From this we deduce the continuity of a homomorphism from a Banach algebra onto a dense subalgebra of a strongly semisimple unital Banach algebra (Rickart [101], Yood [132]).

In Section 7 we prove Kaplansky's Theorem [74] that decomposes a ring isomorphism between two semisimple Banach algebras into a linear part, a conjugate linear part, and a non real linear part on a finite dimensional ideal. This is proved using automatic continuity methods in a similar way to that in which the corresponding result for derivations was proved (Johnson and Sinclair [68]).

In Section 8 we briefly consider the relationship between discontinuous derivations from a Banach algebra A into Banach A-modules and discontinuous derivations from A. From a discontinuous derivation a discontinuous homomorphism may be constructed (Theorem 8.2). Dales's example of a discontinuous derivation from the disc algebra into a Banach module over it is the main result of this section [28]. The structure of the proof given here is slightly different from his but the idea is the same. The existence of a discontinuous homomorphism from the disc algebra into a suitable Banach algebra was first proved using Allan's theorem that embeds the algebra of all formal power series in one indeterminate into suitable Banach algebras [1], (see Johnson [66]). The proof of this excellent deep result is based on several complex variable theory, and was beyond the scope of these lectures.

Section 9 contains the main lemma, Theorem 9. 3, on which Sections 10 and 12 are based. The hypotheses of this theorem have been chosen to suit these two applications.

Section 10 is devoted to Bade and Curtis's theorem on the decomposition of a homomorphism from $\mathbf{C}(\Omega)$, where Ω is a compact Hausdorff space, into continuous and discontinuous parts (Theorem 10. 3) [7]. This is one of the most important results in automatic continuity, and the source of many ideas for subsequent research. A corollary (10. 4) of this theorem is that there is a discontinuous homomorphism from $\mathbf{C}(\Omega)$ into a Banach algebra if and only if there is a μ in Ω and a discontinuous homomorphism from $\mathbf{C}_0(\Omega \setminus \{\mu\})$ into a radical Banach algebra.

In Section 11 properties of a discontinuous homomorphism from $\mathbf{C}_0(\Psi)$ into a radical Banach algebra are studied, where Ψ is a locally compact Hausdorff space. The results of this section depend on Lemma 1. 6 and the observation that positive elements in $\mathbf{C}_0(\Psi)$ have positive roots (Sinclair [118]).

Section 12 contains some results on the continuity of homomorphisms and derivations from C*-algebras. In Corollary 12. 4 we prove that if a unital C*-algebra has no closed cofinite ideals (e. g. $\mathcal{L}(H)$, where H is an infinite dimensional Hilbert space), then each homomorphism from it into a Banach algebra is continuous (Johnson [64]). In Corollary 12. 5 we show that a derivation from a C*-algebra into a Banach bimodule over it is continuous (Ringrose [105] and, see also, Johnson and Parrott [67]).

In Section 13 the standard results on the continuity of positive linear functionals on a Banach $*$-algebra are proved. The automatic continuity of positive linear functionals on other ordered Banach spaces is not considered (see Namioka [94], Peressini [98]). We shall also not consider certain other problems on the automatic continuity of linear functionals. For example, if a linear functional f on a C*-algebra A is continuous on all C*-subalgebras of A generated by single hermitian elements, is f continuous on A (Barnes [13], Barnes and Duncan [14], Ringrose [106])?

These notes are based on a course of postgraduate lectures given at the University of Edinburgh during the spring term 1974. Sections 7 and 13 were not given in the lectures, and Sections 3 and 8 were not covered in detail. I am indebted to those who participated for their suggestions, comments, and perseverance, and to F. F. Bonsall for encouraging me to give the lectures and write the notes. J. Cusack and N. P. Jewell read the manuscript, and their criticism and corrections have prevented many obscurities and errors. I am grateful for their advice. I should like to thank G. R. Allan, J. Cusack, H. G. Dales, B. E. Johnson, T. Lenegan, J. R. Ringrose, and many other friends for discussions, comments, letters, and preprints.

September 1974

1 · Technical results

In this chapter we develop some technical results needed in the subsequent chapters. In Section 1 we study the separating space $\mathfrak{S}(S)$ of a linear operator S from a Banach space X into a Banach space Y, where

$$\mathfrak{S}(S) = \{y \in Y: \text{ there is a sequence } (x_n) \text{ in } X \text{ with } x_n \to 0 \text{ and } Sx_n \to y\}.$$

The separating space is a useful tool in automatic continuity since S is continuous if and only if $\mathfrak{S}(S) = \{0\}$. It has been used by many authors to obtain the continuity of homomorphisms, derivations, module homomorphisms, and intertwining operators (for example [103], [68], [117], [69]). This tradition is followed in these notes. The proof of Lemma 1. 6 illustrates the typical rolling hump argument of automatic continuity proofs.

The main result in Section 2 concerns the continuity behaviour of a linear operator with a large lattice of closed invariant subspaces, with properties akin to the open subsets of a compact Hausdorff space. This method of relating the discontinuity of the linear operator to a finite number of points in an associated topological space occurs in various forms in the following papers: [27], [48], [58], [63], [64], [66], [69], [105], [117], [121], [128]. Theorem 2. 3 does not have as wide an application as we should wish but we are able to apply it later to study the continuity of a linear operator intertwining with a pair of normal operators, to prove the uniqueness of the complete norm topology on a semisimple Banach algebra, and to handle problems concerning additive operators.

1. The separating space

1. 1. **Definition.** If S is a linear operator from a Banach space

X into a Banach space Y, we let $\mathfrak{S}(S)$ or $\mathfrak{S}$ denote the set

$$\{y \in Y\colon \text{there is a sequence } (x_n) \text{ in } X \text{ with } x_n \to 0 \text{ and } Sx_n \to y\},$$

and call it the *separating space* of S.

The first three lemmas contain the elementary properties of the separating space that we shall require in later chapters, and these lemmas will often be used without reference.

1.2. Lemma. *Let* S *be a linear operator from a Banach space* X *into a Banach space* Y. *Then*

(i) $\mathfrak{S}$ *is a closed linear subspace of* Y,

(ii) S *is continuous if and only if* $\mathfrak{S} = \{0\}$, *and*

(iii) *if* T *and* R *are continuous linear operators on* X *and* Y, *respectively, and if* $ST = RS$, *then* $R\mathfrak{S} \subseteq \mathfrak{S}$.

Proof. (i) The separating space is trivially a linear subspace of Y. Let (y_n) be a sequence in $\mathfrak{S}$ with $y_n \to y$ in Y. Choose a sequence (x_n) in X so that $\|x_n\| < 1/n$ and $\|Sx_n - y_n\| < 1/n$ for all n. Then $x_n \to 0$ and $Sx_n \to y$ as n tends to infinity. Hence $\mathfrak{S}$ is closed.

(ii) This is just the closed graph theorem in a different notation. If $\mathfrak{S} = \{0\}$, then S has a closed graph because $x_n \to x$ and $Sx_n \to y$ imply that $x_n - x \to 0$ and $S(x_n - x) \to y - Sx$ so that $y = Sx$.

(iii) If $x_n \to 0$ and $Sx_n \to y$, then $Tx_n \to 0$ and $STx_n = RSx_n \to Ry$.

1.3. Lemma. *Let* S *be a linear operator from a Banach space* X *into a Banach space* Y, *and let* R *be a continuous linear operator from* Y *into a Banach space* Z. *Then*

(i) RS *is continuous if and only if* $R\mathfrak{S}(S) = \{0\}$,

(ii) $(R\mathfrak{S}(S))^{-} = \mathfrak{S}(RS)$, *and*

(iii) *there is a constant* M (*independent of* R *and* Z) *such that if* RS *is continuous then* $\|RS\| \le M\|R\|$.

Proof. (i) If RS is continuous, $x_n \to 0$, and $Sx_n \to y$, then $RSx_n \to Ry$ and $RSx_n \to 0$ so that $Ry = 0$.

Conversely suppose that $R\mathfrak{S}(S) = \{0\}$. The continuity of RS will follow from the commutativity of the diagram

$$\begin{array}{ccccc} & S & & Q & \\ X & \to & Y & \to & Y/\mathfrak{S}(S) \\ & & R \searrow & \swarrow R_0 & \\ & & & Z & \end{array}$$

once we have proved QS continuous; where Q is the natural quotient operator $Y \to Y/\mathfrak{S}(S) : y \mapsto y + \mathfrak{S}(S)$, and $R_0(y + \mathfrak{S}(S)) = Ry$. Let $x_n \to 0$ in X and $QSx_n \to y + \mathfrak{S}(S)$ in $Y/\mathfrak{S}(S)$. Then there is a sequence (y_n) in $\mathfrak{S}(S)$ such that $Sx_n - y - y_n \to 0$. We choose a sequence (w_n) in X so that $\|w_n\| < 1/n$ and $\|Sw_n - y_n\| < 1/n$. Then $x_n - w_n \to 0$ and $S(x_n - w_n) - y \to 0$ as $n \to \infty$ so that y is in $\mathfrak{S}(S)$. Thus $\mathfrak{S}(QS)$ is $\{0\}$ and QS is continuous.

(ii) We have $R\mathfrak{S}(S) \subseteq \mathfrak{S}(RS)$ because $x_n \to 0$ and $Sx_n \to y$ imply that $RSx_n \to Ry$. Since $\mathfrak{S}(RS)$ is closed, it follows that $(R\mathfrak{S}(S))^- \subseteq \mathfrak{S}(RS)$. Let $Q_0 : Z \to Z/(R\mathfrak{S}(S))^- : z \mapsto z + (R\mathfrak{S}(S))^-$. Then $Q_0R\mathfrak{S}(S)$ is null so that Q_0RS is continuous by (i), and thus $Q_0\mathfrak{S}(RS)$ is null also by (i). Therefore

$$\mathfrak{S}(RS) \subseteq (R\mathfrak{S}(S))^-.$$

(iii) Using the proof of (i) and $\|R\| = \|R_0\|$ we obtain

$$\|RS\| = \|R_0QS\| \leq \|QS\|.\ \|R_0\| = \|QS\|.\ \|R\|.$$

Let $M = \|QS\|$, and the proof is complete.

From the above lemma it follows that $S^{-1}\mathfrak{S}(S)$ is closed because it is just $\operatorname{Ker} QS$, where Q is defined as in the proof of (i).

1.4. Lemma. _Let_ X_0 _and_ Y_0 _be closed linear subspaces of Banach spaces_ X _and_ Y, _and let_ S _be a linear operator from_ X _into_ Y _such that_ $SX_0 \subseteq Y_0$. _Let_ $S_0 : X/X_0 \to Y/Y_0$ _be defined by_ $S_0(x + X_0) = Sx + Y_0$. _Then_ S_0 _is continuous if and only if_ $Y_0 \supseteq \mathfrak{S}(S)$.

Proof. If S_0 is continuous, $x_n \to 0$, and $Sx_n \to y$, then $S_0(x_n + X_0)$ tends to Y_0 and to $y + Y_0$ so that $\mathfrak{S}(S) \subseteq Y_0$. Conversely suppose that $\mathfrak{S}(S) \subseteq Y_0$. Let $Q : Y \to Y/Y_0 : y \mapsto y + Y_0$. Then QS is continuous and QS annihilates X_0 so that $S_0(x + X_0) = QS(x)$ and S_0 is continuous.

With the hypotheses of the above lemma we also have $\mathfrak{S}(S|X_0) \subseteq Y_0 \cap \mathfrak{S}(S)$, where $S|X_0$ is the restriction of S to X_0. This inclusion can be strict.

When R is a continuous linear operator, the above lemmas adequately describe the behaviour of the separating space of RS in terms of that of S. This raises the question of how does the separating space of ST behave for T a continuous linear operator from a Banach space into X. The general situation for ST is not as nice as that for RS. Clearly $\mathfrak{S}(ST) \subseteq \mathfrak{S}(S)$ but equality does not hold in general as one can see if S annihilates the range of T. The following result, which is a direct application of the open mapping theorem, is occasionally useful.

1.5. **Lemma.** _Let_ $X, Z_1, \ldots, Z_n$ _be Banach spaces, and let_ $T_1, \ldots, T_n$ _be continuous linear operators from_ $Z_1, \ldots, Z_n$ _into_ X, _respectively, such that_ $X = T_1Z_1 + \ldots + T_nZ_n$. _Let_ S _be a linear operator from_ X _into a Banach space_ Y. _Then_ $\mathfrak{S}(S) = (\mathfrak{S}(ST_1) + \ldots + \mathfrak{S}(ST_n))^-$.

Proof. Suppose that $ST_1, \ldots, ST_n$ are continuous. Let $Z = Z_1 \oplus \ldots \oplus Z_n$ with norm $\|(z_1, \ldots, z_n)\| = \sum_1^n \|z_j\|$, and let $T : Z \to X : (z_1, \ldots, z_n) \mapsto T_1z_1 + \ldots + T_nz_n$. Then T is a continuous linear operator from Z onto X so is an open mapping. Thus there is a constant M such that x in X implies that there is a z in Z with $\|z\| \leq M\|x\|$ and $Tz = x$. The continuity of ST_j for all j gives the continuity of ST. For x in X, and z as above, we have $\|Sx\| = \|STz\| \leq \|ST\|.M.\|x\|$, so S is continuous (see [151]).

We now consider the general case. Since $\mathfrak{S}(ST_j) \subseteq \mathfrak{S}(S)$ for each j we have just to prove that $\mathfrak{S}(S) \subseteq W$, where W is the closure of $\mathfrak{S}(ST_1) + \ldots + \mathfrak{S}(ST_n)$. If Q is the natural quotient operator from Y onto Y/W, then QST_j is continuous for each j so that QS is continuous by the previous paragraph. Hence $\mathfrak{S}(S) \subseteq W$ by Lemma 1.2 and the proof is complete.

The above result does not hold if the set of operators $\{T_j\}$ is infinite. For example take any discontinuous linear operator from a Banach space X into a Banach space Y. Let the Z's run over all finite dimensional subspaces of X, and the T's be the corresponding injections of the Z's into X. Then ST is continuous for all T and the space spanned by the TZ's is X, but S is discontinuous.

The next lemma is the main result of this section and is the crucial lemma on which Sections 4 and 11 are based. It states that a certain descending sequence of closed linear subspaces obtained from S via a countable family of continuous linear operators intertwining with S is eventually constant [118]. It is stronger than Johnson and Sinclair [69, Theorem 3.2] and is related to Allan [1, Theorem 1] (see [152]).

1.6. Lemma. Let X and Y be Banach spaces and let (T_n) and (R_n) be sequences of continuous linear operators on X and Y, respectively. If S is a linear operator from X into Y satisfying $ST_n = R_n S$ for all n, then there is an integer N such that $(R_1 \ldots R_n \mathfrak{S})^- = (R_1 \ldots R_N \mathfrak{S})^-$ for all $n \geq N$.

Proof. Since $ST_{n+1} = R_{n+1}S$ we have $(R_1 \ldots R_n \mathfrak{S})^- \supseteq (R_1 \ldots R_{n+1} \mathfrak{S})^-$ for all n by 1.2. If this inclusion is strict for infinitely many n, then by grouping the R's and T's into finite products corresponding to the intervals of constancy of $(R_1 \ldots R_n \mathfrak{S})^-$ we may assume that

$$(R_1 \ldots R_n \mathfrak{S})^- \supset (R_1 \ldots R_{n+1} \mathfrak{S})^-$$

for all n. We may also assume that $\|T_n\| \leq 1$ for all n.

Let Q_n denote the natural quotient operator from Y onto $Y/(R_1 \ldots R_n \mathfrak{S})^-$ for each n. Then $Q_n R_1 \ldots R_n S$ is continuous and $Q_n R_1 \ldots R_{n-1} S$ is discontinuous for each n. We inductively choose a sequence (x_n) from X so that

$$\|x_n\| \leq 2^{-n}, \text{ and}$$

$$\|Q_n R_1 \ldots R_{n-1} S x_n\| \geq n + \|Q_n R_1 \ldots R_n S\| + \|Q_n S(\sum_1^{n-1} T_1 \ldots T_{j-1} x_j)\|$$

for all n. Let $z = \sum_1^\infty T_1 \dots T_{n-1} x_n$. Then for each positive integer n

$$\begin{aligned} &\|Sz\| \\ &\geq \|Q_n Sz\| \\ &\geq \|Q_n ST_1 \dots T_{n-1} x_n\| - \|Q_n S(\sum_1^{n-1} T_1 \dots T_{j-1} x_j)\| \\ &\quad - \|Q_n ST_1 \dots T_n(\sum_{n+1}^\infty T_{n+1} \dots T_{j-1} x_j)\| \\ &\geq n \end{aligned}$$

by the properties of Q_n and the choice of x_n. This contradiction proves the lemma.

1.7. Corollary. _Let_ X _and_ Y _be Banach spaces, let_ $\mathcal{G}$ _be a commutative semigroup, let_ $\alpha \mapsto T(\alpha)$ _and_ $\alpha \mapsto R(\alpha)$ _be homomorphisms from_ $\mathcal{G}$ _into the multiplicative semigroups of_ $\mathcal{L}(X)$ _and_ $\mathcal{L}(Y)$, _respectively, and let_ S _be a linear operator from_ X _into_ Y _such that_ $ST(\alpha) = R(\alpha)S$ _for all_ α _in_ $\mathcal{G}$. _Then there is a_ γ _in_ $\mathcal{G}$ _such that_ $(R(\alpha)(R(\gamma)\mathfrak{S})^-)^- = (R(\gamma)\mathfrak{S})^-$ _for all_ α _in_ $\mathcal{G}$, _and_ $(R(\gamma)\mathfrak{S})^-$ _contains all closed linear subspaces_ Y _of_ $\mathfrak{S}$ _satisfying_ $(R(\alpha)Y)^- = Y$ _for all_ α _in_ $\mathcal{G}$.

Proof. We begin by observing that it is sufficient to show that there is a γ in $\mathcal{G}$ such that $(R(\gamma)\mathfrak{S})^- \subseteq (R(\alpha)\mathfrak{S})^-$ for all α in $\mathcal{G}$. Because then $(R(\gamma)\mathfrak{S})^- \subseteq (R(\gamma.\alpha)\mathfrak{S})^- = (R(\gamma)R(\alpha)\mathfrak{S})^- \subseteq (R(\gamma)\mathfrak{S})^-$ and $(R(\alpha)R(\gamma)\mathfrak{S})^- = (R(\alpha)(R(\gamma)\mathfrak{S})^-)^-$ for all α in $\mathcal{G}$. Thus $(R(\alpha)(R(\gamma)\mathfrak{S})^-)^- = (R(\gamma)\mathfrak{S})^-$ for all α in $\mathcal{G}$. If Y is a closed linear subspace of $\mathfrak{S}$ and if $(R(\gamma)Y)^- = Y$, then Y is clearly contained in $(R(\gamma)\mathfrak{S})^-$.

Suppose there is no such γ in $\mathcal{G}$. Then for each γ in G there is an α in $\mathcal{G}$ such that $(R(\alpha)\mathfrak{S})^-$ does not contain $(R(\gamma)\mathfrak{S})^-$. Hence $(R(\alpha\gamma)\mathfrak{S})^- \subset (R(\gamma)\mathfrak{S})^-$ because $(R(\alpha\gamma)\mathfrak{S})^- = (R(\alpha)R(\gamma)\mathfrak{S})^- \subseteq (R(\alpha)\mathfrak{S})^-$ and $(R(\alpha\gamma)\mathfrak{S})^- = (R(\gamma)R(\alpha)\mathfrak{S})^- \subseteq (R(\gamma)\mathfrak{S})^-$. We now choose $\alpha_1, \alpha_2, \dots$ in $\mathcal{G}$ by induction so that $(R(\alpha_1 \dots \alpha_n)\mathfrak{S})^- \subset (R(\alpha_1 \dots \alpha_{n-1})\mathfrak{S})^-$ for all n. This contradicts Lemma 1.6 with $T_n = T(\alpha_n)$ and $R_n = R(\alpha_n)$, and completes the proof.

In applying the above lemma and corollary we shall often require conditions which force $\cap (R_1 \ldots R_n \mathfrak{S})^-$ to be null. This will then give us $R_1 \ldots R_N \mathfrak{S} = \{0\}$ for some positive integer N so that $R_1 \ldots R_N S$ is continuous. We require the next lemma when we apply Lemma 1.7 in Section 4. This lemma may be obtained from the Mittag-Leffler Theorem of Bourbaki [19, p. 212] as is noted in Johnson and Sinclair [69, p. 535] and shown in Allan [1]. We give a direct proof of Lemma 1.8. Note that it provides a topological link between subspaces one of which is algebraically maximal and the other topologically maximal.

1.8. **Lemma.** Let (R_n) be a countable commuting sequence of continuous linear operators on a Banach space Y. Let Y_∞ be the maximal linear subspace of Y such that $R_n Y_\infty = Y_\infty$ for all n, and let Y^∞ be the maximal closed linear subspace of Y such that $(R_n Y^{\infty})^- = Y^\infty$ for all n. Then $Y^\infty = (Y_\infty)^-$.

Proof. Clearly $(Y_\infty)^-$ has the property that $(R_n (Y_\infty)^-)^- = (Y_\infty)^-$ for all n so that $(Y_\infty)^- \subseteq Y^\infty$. We have to prove that Y_∞ is dense in Y^∞. Renumber the sequence (R_n) so that each R_n occurs infinitely often in the new sequence (R_n). We may assume that $\|R_n\| = 1$ for all n.

Let y be in Y^∞ and let $\varepsilon > 0$. Since $(R_n Y^{\infty})^- = Y^\infty$ for all n we choose a sequence (x_n) from Y^∞ so that $x_0 = y$, $\|x_n - R_{n+1} x_{n+1}\| < \varepsilon . 2^{-n-1}$ for all n. Then for m, n, p non-negative integers with $m < n$ we have

$$
\begin{aligned}
&\|R_p \ldots R_{p+m} x_{p+m} - R_p \ldots R_{p+n} x_{p+n}\| \\
&\leq \sum_{j=m}^{n-1} \|R_p \ldots R_{p+j} x_{p+j} - R_p \ldots R_{p+j+1} x_{p+j+1}\| \\
&\leq \sum_{j=m}^{n-1} \|x_{p+j} - R_{p+j+1} x_{p+j+1}\| \\
&< \sum_{j=m}^{n-1} \varepsilon . 2^{-p-j-1} < \varepsilon . 2^{-p-m}. \qquad (\dagger)
\end{aligned}
$$

Thus the sequence $(R_p \ldots R_{p+m} x_{p+m} : m = 1, 2, \ldots)$ is Cauchy in Y^∞ for each p. We denote the limit of this sequence by y_p. Then

$R_{p-1}y_p = y_{p-1}$ for all positive integers p.

Taking $p = m = 0$ in (†) and noting that we have assumed that $R_0 = I$ we obtain $\|x_0 - y_0\| \le \varepsilon$. Let Z be the linear space generated by $R_1^{\mu_1} \dots R_n^{\mu_n} y_j$ for all positive integers j, and all finite sets of non-negative integers $\{\mu_1, \dots, \mu_n\}$. Then $R_m Z \subseteq Z$ for all m. Let m and j be positive integers. Then there is a k such that $R_k = R_m$ and $k > j$ because each element appears infinitely often in the sequence (R_n). Therefore $y_j = R_{j+1} \dots R_k y_{k+1} = R_m(R_{j+1} \dots R_{k-1} y_{k+1})$ so that y_j is in $R_m Z$. Thus $R_m Z = Z$ for all m. The maximality of Y_∞ implies that $Z \subseteq Y_\infty$, and the proof is complete.

The above lemma enables us to strengthen 1.7 for countable semigroups.

1.9. Corollary. *Let X and Y be Banach spaces, let $\mathcal{G}$ be a countable commutative semigroup, and let $\alpha \mapsto T(\alpha)$ and $\alpha \mapsto R(\alpha)$ be homomorphisms from $\mathcal{G}$ into $\mathcal{L}(X)$ and $\mathcal{L}(Y)$, respectively. Let $\mathfrak{S}_\infty$ be the maximal linear subspace of $\mathfrak{S}$ such that $R(\alpha)\mathfrak{S}_\infty = \mathfrak{S}_\infty$ for all α in $\mathcal{G}$. Then there is a γ in $\mathcal{G}$ such that $(R(\gamma)\mathfrak{S})^- = (\mathfrak{S}_\infty)^-$.*

2. Discontinuity points

The following elementary consequence of the open mapping theorem is used in the proof of Theorem 2.3. It occurs in Johnson [59], and we give it here to establish notation.

2.1. Lemma. *Let X be a Banach space and let V and W be closed linear subspaces of X such that $X = V + W$. Then there is a real number k such that for each x in X there are v in V and w in W with $x = v + w$ and $\|v\| + \|w\| \le k\|x\|$.*

Proof. Apply the open mapping theorem to the continuous linear operator $\psi : V \oplus W \to X : (v, w) \mapsto v + w$, where $\|(v, w)\| = \|v\| + \|w\|$.

2.2. Conditions. We now introduce the notation and definitions that will apply throughout this section and that will be referred to in the various applications of these results. Let X be a Banach space, let Y

be a normed space, let Ω be a regular (Hausdorff) topological space, and let $F \mapsto X(F)$ and $F \mapsto Y(F)$ be maps from a basis Γ of the topology on Ω into the set of closed linear subspaces of X such that if $F_1, \ldots, F_n$ are in Γ with $F_j^- \cap F_h^- = \emptyset$ for $j \neq h$, then

$$X(F_1) \cap \ldots \cap X(F_{n-1}) + X(F_n) = X.$$

Let $Q(F)$ be the natural quotient operator from Y onto $Y/Y(F)$. If S is a linear operator from X into Y, a point λ in Ω is called a discontinuity point of S (with respect to $F \mapsto Y(F)$) if $Q(F)S$ is discontinuous for all F in Γ with λ in F.

2.3. Theorem. Assume that the Conditions 2.2 hold. If S is a linear operator from X into Y satisfying $SX(F) \subseteq Y(F)$ for all F in Γ, then S has only a finite number of discontinuity points.

Proof. Suppose that S has an infinite number of discontinuity points. We shall choose two sequences (U_n) and (V_n) of open subsets in Ω such that $U_n \cap V_n = \emptyset$, U_n contains a discontinuity point of S, and $U_n \subseteq V_j$ for $1 \leq j \leq n-1$. To ensure that the induction can proceed we also require that $V_1 \cap \ldots \cap V_n$ contains infinitely many discontinuity points for all n. Choose U_1, V_1 disjoint open sets so that U_1 contains at least one discontinuity point, V_1 contains an infinite number of discontinuity points, and $U_1 \cap V_1 = \emptyset$. Suppose $U_1, \ldots, U_n$ and $V_1, \ldots, V_n$ have been chosen. We now choose disjoint open subsets U'_{n+1} and V_{n+1} so that U'_{n+1} contains at least one discontinuity point that is in $V_1 \cap \ldots \cap V_n$ and V_{n+1} contains an infinite number of discontinuity points in $V_1 \cap \ldots \cap V_n$. Let $U_{n+1} = U'_{n+1} \cap V_1 \cap \ldots \cap V_n$. This completes the inductive choice of (U_n) and (V_n).

Using the regularity of Ω, for each n we choose an open subset W_n in Ω such that W_n contains a discontinuity point λ_n and $W_n \subseteq W_n^- \subseteq U_n$. We now choose an F_n in Γ so that $\lambda_n \in F_n \subseteq W_n$. Thus $F_n^- \subseteq W_n^- \subseteq U_n$. If $m > n$, then $F_n^- \cap F_m^- \subseteq U_n \cap U_m \subseteq U_n \cap V_n = \emptyset$ by the construction of U_m, U_n, and V_n. Because $F_m^- \cap F_n^- = \emptyset$ if $m \neq n$, we have $X(F_1) \cap \ldots \cap X(F_{n-1}) + X(F_n) = X$ for each positive integer n by 2.2. Let k_n be the positive real number given by Lemma

2.1 corresponding to this decomposition of X as a sum of two closed linear subspaces.

By induction on n we choose two sequences (x_n) and (z_n) in X such that

(1) $\|x_n\| < 2^{-n}k_n^{-1}$,

(2) $\|Q(F_n)Sx_n\| \geq n + \|S(\sum_1^{n-1} z_j)\|$,

(3) $x_n \in X(F_1) \cap \ldots \cap X(F_{n-1})$,

(4) $x_n - z_n \in X(F_n)$, and

(5) $\|z_n\| \leq k_n \|x_n\|$

hold for all $n \geq 2$. Let x_1 be any element of X and let $z_1 = 0$. Suppose that both sequences have been chosen to the $(n-1)^{th}$-element. The discontinuity of $Q(F_n)S$ enables us to choose an x_n in X satisfying (1) and (2). By Lemma 2.1 we may now choose a z_n in X satisfying (3), (4), and (5). We let $z = \sum_1^{\infty} z_j$ observing that the series converges by (1) and (5). Since $X(F_n)$ is a closed linear subspace of X, property (3) implies that $\sum_{j>n} z_j$ is in $X(F_n)$ for all n. Hence $S(\sum_{j>n} z_j)$ is in $Y(F_n)$ by hypothesis, so that $Q(F_n)S(\sum_{j>n} z_j) = 0$. Similarly $Q(F_n)Sz_n = Q(F_n)Sx_n$ for all n. Therefore

$$\begin{aligned} \|Sz\| &\geq \|Q(F_n)Sz\| \\ &\geq \|Q(F_n)Sz_n\| - \|S(\sum_{j<n} z_j)\| \\ &= \|Q(F_n)Sx_n\| - \|S(\sum_{j<n} z_j)\| \\ &\geq n \end{aligned}$$

for all positive integers n. This contradiction completes the proof.

2.4. Remarks. (a) The hypothesis $SX(F) \subseteq Y(F)$ for all F in Γ in Theorem 2.3 may be altered to the assumption that $SX(F) \subseteq Y(G)$ if F and G are in Γ and satisfy $G^- \subseteq F$. The proof requires minor alterations. After choosing F_n we choose an open subset H_n of Ω such that $\lambda_n \in H_n \subseteq H_n^- \subseteq F_n$. We then choose a G_n in Γ so that

$\lambda_n \in G_n \subseteq H_n$. The remainder of the proof is similar with some of the F_n replaced by G_n.

(b) By imposing various restrictions on the subspaces $Y(F)$ we can deduce the continuity of S on certain subspaces of X, and the continuity of $Q(F)S$ for certain F in Γ.

2.5. Corollary. *Let* Ω *be a compact Hausdorff space, let* Y *be a Banach space, and let* F *be the finite set of discontinuity points of* S *given by Theorem 2.3.*

(i) *If the map* $G \mapsto Y(G)$ *satisfies the condition that* $V_1, \ldots, V_n$ *in* Γ *with* $\bigcup_1^n V_j = \Omega$ *implies that* $\bigcap_1^n Y(V_j) = \{0\}$, *then for each finite set* $\{W_1, \ldots, W_n\}$ *in* Γ *with* $F \subseteq W_1 \cup \ldots \cup W_n$ *the operator* S *is continuous when restricted to* $X(W_1) \cap \ldots \cap X(W_n)$.

(ii) *If* $V, V_1, \ldots, V_n$ *in* Γ *with* $V^- \subseteq \bigcup_1^n V_j$ *implies that* $Y(V) \supseteq \bigcap_1^n Y(V_j)$, *then* $Q(U)S$ *is continuous for all* U *in* Γ *such that* $U^- \cap F = \emptyset$.

Proof. (i) The set $\Omega \backslash \bigcup_1^n W_j$ is compact and for each λ in this set there is a U in Γ containing λ with $Q(U)S$ continuous (2.2). Thus there is a finite subset $\{U_1, \ldots, U_n\}$ of Γ such that $\bigcup_1^m U_j \cup \bigcup_1^n W_k = \Omega$ and $Q(U_h)S$ is continuous for $h = 1, \ldots, m$. By hypothesis

$$Y(U_1) \cap \ldots \cap Y(U_m) \cap Y(W_1) \cap \ldots \cap Y(W_n) = \{0\}.$$

Since $\mathfrak{S}(S) \subseteq Y(U_j)$ for each j, we have

$$\mathfrak{S}(S) \cap Y(W_1) \cap \ldots \cap Y(W_n) = \{0\}.$$

Since $S(\bigcap_1^n X(W_j)) \subseteq \bigcap_1^n Y(W_j)$, the separating space of S restricted to $\bigcap_1^n X(W_j)$ is contained in $\bigcap_1^n Y(W_j)$. Hence the separating space of the restriction of S to $\bigcap_1^n X(W_j)$ is zero, and so the restriction is continuous.

(ii) To each λ in U^- there corresponds an element G of Γ such that λ is in G and $Q(G)S$ is continuous. Because U^- is compact there are a finite number of elements $G_1, \ldots, G_n$ of Γ such that $U^- \subseteq G_1 \cup \ldots \cup G_n$. By hypothesis $Y(U) \supseteq Y(G_1) \cap \ldots \cap Y(G_n)$. Because $Q(G_j)S$ is continuous for each j, the separating space of S is contained in $Y(G_j)$, and so $\mathfrak{S}(S)$ is contained in $Y(U)$. Therefore $Q(U)S$ is continuous and the proof is complete.

2. 6. Remarks. One of the aims of this section was to develop a standard technical lemma sufficient for all important automatic continuity results that use some form of 'discontinuity point'. We have not succeeded in this objective because Theorem 2. 3 is not strong enough to give the results of Bade and Curtis on the continuity of S on a large subspace of X, [7] (§10). A disadvantage of Theorem 2. 3 is that hypothesis 2. 2(i) is on the domain space and the conclusions are in the range space. Corollary 2. 5(i) is an attempt to get round this difficulty by ensuring that the conclusion concerns the domain space, but to obtain this result we make an assumption on the range space. An alternative approach here is to call λ in Ω a range space discontinuity point if the restriction of S to $X(F)$ is discontinuous for all F containing λ. This type of discontinuity point is used by P. Vrbová [128, p. 143] to study the continuity of a linear operator intertwining with a pair of generalized scalar operators.

I do not know if the continuity of derivations on a semisimple Banach algebra [68] can be obtained from Theorem 2. 3 or something like it. Possibly Theorem 2. 3 is moving in the wrong direction from the uniqueness of the complete norm topology theorem [59] (§6) and a result like Pták [99] would actually be more useful. The difficulty is to find a natural example to which Pták's theorem will apply other than the uniqueness of the complete norm topology on a semisimple Banach algebra. A nice feature of [99] is its close link with multilinear operators. A more hopeful group of theorems exploiting multilinear operators are those of Laursen and Stein [78], and Stein [141].

2 · Intertwining operators

In this chapter we discuss some necessary and sufficient conditions on a pair (T, R) of continuous linear operators on Banach spaces X and Y, respectively, so that each linear operator S from X into Y intertwining with T and R (i.e. satisfying $ST = RS$) is continuous. We begin with two results implying the existence of discontinuous intertwining operators (Section 3). The first requires the existence of a complex number μ such that μ is an eigenvalue of R and $(T - \mu I)X$ is of infinite codimension in X. The second depends on there being a non-zero linear subspace Z of Y such that $(R - \mu I)Z = Z$ for all complex numbers μ, and on T not being algebraic.

In Section 4 we give necessary and sufficient conditions for each linear operator S intertwining with the pair (T, R) to be continuous when the spectrum of R is countable, and in Section 5 we consider the case when T and R are normal operators on a Hilbert space.

3. The existence of discontinuous intertwining operators

Throughout this section X and Y are Banach spaces, and T and R are continuous linear operators on X and Y, respectively.

3.1. **Definition.** A complex number μ is said to be a critical eigenvalue of the pair (T, R) if $(T - \mu I)X$ is of infinite codimension in X and μ is an eigenvalue of R.

3.2. **Lemma.** *If (T, R) has a critical eigenvalue, then there is a discontinuous linear operator S from X into Y intertwining with (T, R).*

Proof. Let μ be a critical eigenvalue of (T, R). Since $X/(T - \mu I)X$ is of infinite dimension, we can use Zorn's lemma to choose a discontinuous linear functional f on X such that $f((T - \mu I)X) = \{0\}$.

Let $y \neq 0$ be a μ-eigenvector of R in Y, and let S be defined by $S(x) = f(x)y$ for all x in X. Then $S(T - \mu I) = 0 = (R - \mu I)S$, which completes the proof of the lemma.

The following lemma shows why, in the previous lemma, we do not have to consider the case when $(T - \mu I)X$ is of finite codimension but is not closed. When $(T - \mu I)X$ is not closed there is a discontinuous linear functional on X annihilating $(T - \mu I)X$, and the proof of Lemma 3.2 applies. We shall use Lemma 3.3 in Section 4.

3.3. **Lemma.** _Let T be a continuous linear operator from a Banach space X into a Banach space Z. If TX is of finite codimension in Z, then TX is closed in Z._

Proof. Let W be a finite dimensional linear subspace of Z such that $TX \oplus W = Z$ as linear spaces. We define $\psi : X \oplus W \to Z$ by $\psi(x, w) = Tx + w$, and give $X \oplus W$ the norm $\|(x, w)\|_1 = \|x\| + \|w\|$. Since X and W are Banach spaces, $(X \oplus W, \|.\|_1)$ is a Banach space. Further ψ is a continuous linear operator from $X \oplus W$ onto Z so ψ is an open mapping. The set $\{(x, w) : w \neq 0\}$ is open in $X \oplus W$ and thus $\{Tx + w : x \in X, w \neq 0\}$ is open in Z. Therefore TX is closed.

3.4. **Remarks.** Let $C[\chi]$ denote the ring of polynomials with complex coefficients in an indeterminate χ. We regard X and Y as $C[\chi]$-modules via the linear operators T and R, i.e., if p is in $C[\chi]$ we define p.x and p.y by $p.x = p(T)x$ and $p.y = p(R)y$. Then S is a module homomorphism from X into Y if and only if S is a linear operator from X into Y satisfying $ST = RS$. A linear subspace Z of Y is said to be R-_divisible_ if $(R - \mu I)Z = Z$ for all complex numbers μ. Note that an R-divisible subspace of Y is closed if and only if it is zero [69]. This is because if Z is a non-zero closed linear space and μ is in the boundary of the spectrum of R restricted to Z then $(R - \mu I)Z \neq Z$. By the fundamental theorem of algebra elements in $C[\chi]$ may be factorized into linear factors. Thus Z is R-divisible if and only if $p.Z = Z$ for all non-zero p in $C[\chi]$, and so Z is R-divisible if and only if Z is divisible as a $C[\chi]$-module. The ring $C[\chi]$ is a principal ideal domain. Hence a $C[\chi]$-module is divisible if and only if it is injective (Cartan and Eilenberg [22, p. 134], Kaplansky [138]). In this section all our modules will be over $\mathbf{C}[\chi]$. Recall that a module Z is <u>injective</u> if for each submodule M of a module W and each homomorphism S_0 from M into Z there is a homomorphism from W into Z whose restriction to M is S_0.

We now give the second method of obtaining a discontinuous intertwining operator by constructing a discontinuous homomorphism from a submodule of X into an injective submodule of Y. The injective property enables us to lift this homomorphism to a homomorphism from X into Y. In Lemma 3.5 we shall obtain the submodule of X on which the initial homomorphism is defined, and in Theorem 3.6 we complete the proof.

An operator T on a space X is called <u>algebraic</u> if there is a non-zero polynomial p such that $p(T) = 0$. A $\mathbf{C}[\chi]$-module X is said to have <u>an infinitely generated free submodule</u> Z if there is an infinite set U in X such that Z is the algebraic direct sum of $\mathbf{C}[\chi].u$ as u runs over U, and each $\mathbf{C}[\chi].u$ is $\mathbf{C}[\chi]$-isomorphic to $\mathbf{C}[\chi]$ under the map $p(\chi).u \mapsto p(\chi)$.

3.5. **Lemma.** <u>If</u> T <u>is not algebraic, then</u> X <u>contains an infinitely generated free</u> $\mathbf{C}[\chi]$-<u>submodule.</u>

Proof. Let X^ω be the product of countably many copies of X with the product topology, and let $\mathbf{C}[\chi]_0$ be the space of all sequences from $\mathbf{C}[\chi]$ with only a finite number of non-zero entries. If $P = (p_j)$ is in $\mathbf{C}[\chi]_0$ and $y = (y_j)$ is in X^ω, then we define $Py = \Sigma p_j y_j = \Sigma p_j(T) y_j$. The sum converges as it is actually only a finite sum. For P in $\mathbf{C}[\chi]_0$ we let

$$\deg P = \text{maximum of the degrees of } p_j,$$

$$\operatorname{Supp} P = \{j : p_j \neq 0\}, \text{ and}$$

$$\|P\| = \Sigma_{j,k} |\alpha_{jk}|,$$

where $P = (p_j)$ and $p_j(\chi) = \Sigma_k \alpha_{jk} \chi^k$. For each positive integer n we let

$$F_n = \{y \in X^\omega : P.y = 0 \text{ for some } P \in \mathbf{C}[\chi]_0 \text{ with } \deg P < n,$$
$$\operatorname{Supp} P \subseteq \{1, \ldots, n\}, \text{ and } 2^{-n} \le \|P\| \le 2^n\}.$$

Then F_n is a closed subset of X^ω for each positive integer n because of the compactness of the set of P in $\mathbf{C}[\chi]_0$ that defines F_n.

Suppose that X contains no infinitely generated free $\mathbf{C}[\chi]$-submodule. Then for each infinite sequence (x_n) from X there is a P in $\mathbf{C}[\chi]_0$ such that $P.(x_n) = 0$. Hence the sequence (F_n) of sets covers X^ω. By the Baire Category Theorem there is a positive integer n such that F_n has non-empty interior [39]. Let $z = (z_1, z_2, \ldots)$ be in the interior of F_n, and let W be the linear space generated by $p.z_j$ for $j = 1, 2, \ldots, n$ and p a polynomial of degree less than n. Then the dimension of W is less than or equal to n^2. If $x = (x_1, x_2, \ldots)$ is in X^ω then there is a positive real number β such that $z + \beta x$ is in F_n, and hence there are polynomials $p_1, \ldots, p_n$ each of degree less than n and at least one non-zero such that

$$p_1.(z_1 + \beta x_1) + \ldots + p_n.(z_n + \beta x_n) = 0.$$

Thus $p_1.x_1 + \ldots + p_n.x_n$ is in W. Applying this to the sequence $(T^k x, T^{k+n}x, T^{k+2n}x, \ldots)$ we obtain a non-zero polynomial q_k (with $q_k(\chi) = p_1(\chi)\chi^k + \ldots + p_n(\chi).\chi^{k+(n-1)n}$) such that $q_k.x$ is in W, and the only non-zero coefficients of χ in q_k lie between the k and $(k + n^2 - 1)$. Since the dimension of W is less than or equal to n^2, the set

$$\{q_0.x, q_{n^2}.x, q_{2n^2}.x, \ldots, q_{n^2(n^2+1)}.x\}$$

is linearly dependent, and hence there is a non-zero polynomial r with $\deg r < n^2(n^2+1) + n^2 - 1 = m$ such that $r.x = r(T)x = 0$.

The proof is now completed by a standard algebraic result (see Kaplansky [138]) whose proof we incorporate into this proof. We decompose X into its primary submodules using the standard theory of torsion modules over a principal ideal domain (e.g. [144]). For each α in $\mathbf{C}$, let

$$X_\alpha = \{x \in X : (\alpha - T)^n x = 0 \text{ for some positive integer } n\}.$$

Then X_α is a primary $\mathbf{C}[\chi]$-submodule of X, and X is the direct sum of the X_α for all α in $\mathbf{C}$ - the latter is the important point [138]. Now X_α is non-zero for at most m distinct values of α. Because if $k > m$ and (x_j) is a sequence of non-zero elements from k distinct

X_α, then a non-zero polynomial p annihilates $x_1 + \ldots + x_k$ only if the degree of p is not less than k. This contradicts what we proved in the last paragraph. If $\alpha_1, \ldots, \alpha_m$ are the complex numbers for which X_α is non-zero, and if $p(\chi) = \{(\chi - \alpha_1) \ldots (\chi - \alpha_m)\}^m$, then $p(T) = 0$ since $X = X_{\alpha_1} \oplus \ldots \oplus X_{\alpha_m}$. This contradicts the hypothesis that T is not algebraic and proves the lemma.

3.6. **Theorem.** *Let X and Y be Banach spaces, and let T and R be continuous linear operators on X and Y, respectively. If T is not algebraic, and if Y has a non-null R-divisible subspace, then there is a discontinuous linear operator S from X into Y satisfying ST = RS.*

Proof. By Lemma 3.5 there are $x_1, x_2, \ldots$ in X such that $W = \mathbf{C}[\chi]x_1 \oplus \mathbf{C}[\chi]x_2 \oplus \ldots$ is a direct sum of $\mathbf{C}[\chi]$-submodules of X each isomorphic to $\mathbf{C}[\chi]$. By normalizing x_j we may assume that $\|x_j\| = 1$ for all j. Let y be a non-zero element in an R-divisible subspace of Y. We define $S_0 : W \to Y$ by $S_0 p(T)x_j = jp(R)y$ for each p in $\mathbf{C}[\chi]$ and each j, and by linearity on all of W. Then S_0 maps W into an R-divisible subspace of Y, $S_0 T = RS_0$ on W, and S_0 is discontinuous on W. Hence S_0 is a $\mathbf{C}[\chi]$-homomorphism from a submodule of X into an injective submodule of Y, so S_0 may be extended to a homomorphism S from X into the injective submodule of Y. This S is the required linear operator.

3.7. **Remarks.** (a) Investigating the R-divisible subspaces of a Banach space shows that a slight alteration in the construction of S_0 gives S_0 to be an isomorphism. This implies that the operator S has a relatively large range, and is more complicated than the operator defined in Lemma 3.2 (see Sinclair [116]).

(b) Lemma 3.2 is taken from Johnson [58]. Theorem 3.6 occurs in Sinclair [116] and is a strengthening of a result in Johnson and Sinclair [69]. This method of using divisible subspaces arose from S. Swierczkowski's proof [69, p. 533] that there is a discontinuous linear operator S commuting with the left shift T on $l_2(0, \infty)$ although the pair (T, T) has no critical eigenvalues. S. Grabiner has proved a weaker form of Theorem 3.6 [47].

4. When R has countable spectrum

If R has countable spectrum or if $\sigma(R) \cap \sigma(T)$ is empty, then the negation of the conditions of Section 3 are necessary and sufficient for every linear operator intertwining with (T, R) to be continuous. We prove this in Theorems 4.1 and 4.2.

4.1. Theorem. *Let X and Y be Banach spaces and let T and R be continuous linear operators on X and Y, respectively. Suppose that the spectrum of R is countable. Then every linear operator S from X into Y satisfying $ST = RS$ is continuous if and only if*

(i) *(T, R) has no critical eigenvalue, and*

(ii) *either (a) T is algebraic, or*

(b) $\{0\}$ is the only R-divisible subspace of Y.

Proof. By the results of Section 3 we have only to prove that (i) and (ii) imply that S is continuous. Assume that (i) and (ii)(b) hold. Let $\sigma(R) = \{\lambda_1, \lambda_2, \dots\}$, let $R_n = \lambda_n I - R$, and let $T_n = \lambda_n I - T$. Let $\mathcal{G}$ be the free commutative semigroup generated by a countable number of symbols, and define homomorphisms from $\mathcal{G}$ into $\mathcal{L}(X)$ and $\mathcal{L}(Y)$ by sending the n-th generator of $\mathcal{G}$ to T_n and R_n, respectively. By Corollary 1.8 there is a non-zero polynomial p in R with zeros in $\sigma(R)$ such that $(p(R)\mathfrak{S})^- = (\mathfrak{S}_\infty)^-$, where $\mathfrak{S}_\infty$ is the maximal linear subspace of $\mathfrak{S}$ such that $R_n \mathfrak{S}_\infty = \mathfrak{S}_\infty$ for all n. Clearly $\mathfrak{S}_\infty \subseteq Y_\infty$, where Y_∞ is the maximal linear subspace of Y such that $R_n Y_\infty = (\lambda_n I - R)Y_\infty = Y_\infty$ for all n. If $\lambda \notin \sigma(R)$, then $R_n(\lambda I - R)^{-1}Y_\infty = (\lambda I - R)^{-1}R_n Y_\infty = (\lambda I - R)^{-1}Y_\infty$ so that $(\lambda I - R)^{-1}Y_\infty \subseteq Y_\infty$. This together with $(\lambda I - R)Y_\infty \subseteq Y_\infty$ shows that $(\lambda I - R)Y_\infty = Y_\infty$. Therefore Y_∞ is an R-divisible subspace of Y so that $Y_\infty = \{0\}$, and hence $p(R)\mathfrak{S} = \{0\}$.

By cancelling those factors $\lambda_n I - R$ from p for which $\lambda_n I - R$ is one-to-one, we may assume that all the zeros of p are eigenvalues of R. As (T, R) has no critical eigenvalue it follows that $(\mu I - T)X$ has finite codimension in X for all zeros μ of p. Thus $p(T)X$ has finite codimension in X so that $p(T)X$ is closed (3.3) and $p(T)$ is an open mapping from X onto $p(T)X$. Because $p(R)\mathfrak{S} = \{0\}$, the operator

$p(R)S = Sp(T)$ is continuous. Hence S is continuous on $p(T)X$. Since $p(T)X$ is of finite codimension in X, the operator S is continuous on X.

We now assume that (i) and (ii)(a) hold. Let $\sigma(T) = \{\lambda_1, \ldots, \lambda_n\}$. Because T satisfies a polynomial equation and has spectrum $\{\lambda_1, \ldots, \lambda_n\}$, the annihilator of the $\mathbf{C}[\chi]$-module X in $\mathbf{C}[\chi]$ is the principal ideal

$$(\chi - \lambda_1)^{r_1} \ldots (\chi - \lambda_n)^{r_n}\mathbf{C}[\chi]$$

for some sequence of positive integers $r_1, \ldots, r_n$. Decomposing X into its primary components, the $\mathbf{C}[\chi]$-submodules

$$X_j = \{x \in X : (\chi - \lambda_j)^{r_j}.x = 0\}$$

in X satisfy $X = X_1 \oplus \ldots \oplus X_n$ (see Kaplansky [138], or Hartley and Hawkes [144]). This decomposition may also be obtained by the functional calculus for T. Thus each X_j is a closed linear subspace of X, and the direct sum $X = X_1 \oplus \ldots \oplus X_n$ is a topological direct sum. If λ_j is not an eigenvalue of R, then

$$(R - \lambda_j I)^{r_j} SX_j = S(T - \lambda_j I)^{r_j} X_j = \{0\}$$

so $SX_j = \{0\}$. If λ_j is an eigenvalue of R, then $(T - \lambda_j I)X$ is of finite codimension in X since (T, R) has no critical eigenvalue; hence $(T - \lambda_j I)^{r_j}X$ is of finite codimension in X. If $k \neq j$, then $(T - \lambda_j I)^{r_j}X_k = X_k$. Therefore we have

$$(T - \lambda_j I)^{r_j}X = X_1 \oplus \ldots \oplus X_{j-1} \oplus X_{j+1} \oplus \ldots \oplus X_n.$$

So if λ_j is an eigenvalue of R, the linear space X_j is finite dimensional. Therefore S is zero on the direct sum of the X_k for those λ_k that are not eigenvalues of R. This direct sum is a closed linear subspace of finite codimension in X. Hence S is continuous.

4.2. Theorem. Let X and Y be Banach spaces, and let T and R be continuous linear operators on X and Y, respectively, such that $\sigma(T) \cap \sigma(R)$ is empty. Then every linear operator S from X into

Y satisfying $ST = RS$ is continuous if and only if T is algebraic or $\{0\}$ is the only R-divisible subspace of Y. If these conditions apply, $S = 0$.

Proof. If T is not algebraic and Y has a non-null R-divisible subspace, then Section 3 gives a discontinuous S. Suppose T is algebraic. Let p be a polynomial with zeros in $\sigma(T)$ such that $p(T) = 0$. Then $p(R)S = Sp(T) = 0$, so that $S = 0$ because the zeros of p do not lie in $\sigma(R)$.

Now suppose that $\{0\}$ is the only R-divisible subspace of Y. If λ is in $\sigma(R)$, then $(R - \lambda I)SX = S(T - \lambda I)X = SX$. Let Y_∞ be the maximal linear subspace of Y such that $(R - \lambda I)Y_\infty = Y_\infty$ for all λ in $\sigma(R)$. Then SX is contained in Y_∞. If μ is not in $\sigma(R)$, then $(R - \lambda I)(R - \mu I)^{-1}Y_\infty = (R - \mu I)^{-1}Y_\infty$ for all λ in $\sigma(R)$ so maximality of Y_∞ implies that $(R - \mu I)^{-1}Y_\infty \subseteq Y_\infty$. As in the proof of 4.1 this gives $(R - \mu I)Y_\infty = Y_\infty$, and hence $Y_\infty = \{0\}$. Thus $S = 0$.

4.3. **Remarks.** (a) Theorem 4.2 shows that the hypothesis of continuity cannot be omitted from the well known application of the analytic functional calculus that if (in addition to the hypotheses of Theorem 4.2) S is continuous, then S is zero. The proof of Theorem 4.2 is standard algebra, and does not require the results of Chapter 1.

(b) Can the hypothesis 'the spectrum of R is countable' in Theorem 4.1 be weakened to '$\sigma(R) \cap \sigma(T)$ is at most countable'? This would generalize 4.1 and 4.2, and would seem to be the limit of these methods for a pair of operators (T, R) (see Johnson and Sinclair [69], and Sinclair [116]). The hypothesis of countability is the difficult one to remove but it may be replaced by other hypotheses (see, for example, Section 5).

(c) As Corollary 1.8 holds for a countable commutative semigroup one can obtain the continuity of a linear operator S that intertwines with a suitable countable commuting family in the case corresponding to 4.1(i) and (ii)(b).

(d) If in Theorem 4.1 we know that T is quasinilpotent or compact, then the hypothesis that (T, R) has no critical eigenvalue is equivalent to assuming that R is one-to-one (see Johnson [142]).

5. Operators intertwining with normal operators

In Lemma 5.1 we show that a normal operator on a Hilbert space has $\{0\}$ as the only divisible subspace. This emphasises the relationship of this section with the previous one.

5.1. Lemma. Let T be a normal operator on a Hilbert space H. Then $\cap\{(\lambda I - T)H : \lambda \in \sigma(T)\}$ is zero.

Proof. Let M be a square containing $\sigma(T)$. Then

$$\cap\{(\lambda I - T)H : \lambda \in M\} = \cap\{(\lambda I - T)H : \lambda \in \sigma(T)\}$$

because $(\lambda I - T)H = H$ for all $\lambda \notin \sigma(T)$. Let x be in $\cap\{(\lambda I - T)H : \lambda \in M\}$, and let $E(\cdot)$ be the spectral measure of T. We shall construct a sequence $\{M_n\}$ of Borel subsets of M with the following properties:

(i) $M_0 = M$,

(ii) M_n^- is one of the four closed squares obtained by halving the sides of M_{n-1}^-, and

(iii) $\|E(M)x\|^2 \leq 4^n \|E(M_n)x\|^2$.

Suppose that M_n has been constructed. Let C_1, C_2, C_3, C_4 be four disjoint Borel subsets of M_n such that the closures of the C's are the four squares obtained by bisecting the sides of M_n^- and joining opposite points of bisection, and that $M_n = C_1 \cup C_2 \cup C_3 \cup C_4$. Then $\|E(M_n)x\|^2 = \sum_1^4 \|E(C_j)x\|^2$ by the properties of the spectral measure. So there is a j so that $\|E(M_n)x\|^2 \leq 4\|E(C_j)x\|^2$. Let M_{n+1} be this C_j.

By Cantor's Intersection Theorem there is a λ_0 such that $\bigcap_0^\infty M_n^- = \{\lambda_0\}$. Since λ_0 is in M, there is a y in H with $(\lambda_0 I - T)y = x$. Hence

$$\begin{aligned}\|E(M)x\|^2 &= \|x\|^2 \\ &\leq 4^n \|E(M_n)x\|^2 \\ &= 4^n \|E(M_n)(\lambda_0 I - T)y\|^2\end{aligned}$$

$$= 4^n\langle E(M_n)(\lambda_0 I - T)^*(\lambda_0 I - T)y, y\rangle$$
$$= 4^n \int_{M_n} |\lambda_0 - \lambda|^2 d\langle E(\lambda)y, y\rangle$$
$$= 4^n \int_{M_n\setminus\{\lambda_0\}} |\lambda_0 - \lambda|^2 d\langle E(\lambda)y, y\rangle$$

for all positive integers n. Therefore

$$\|x\|^2 \le 4^n(\text{diam } M_n)^2 \langle E(M_n\setminus\{\lambda_0\})y, y\rangle$$
$$\le (\text{diam } M)^2 \langle E(M_n\setminus\{\lambda_0\})y, y\rangle .$$

Now $\langle E(M_n\setminus\{\lambda_0\})y, y\rangle \to 0$ as $n \to \infty$, because $\langle E(\cdot)y, y\rangle$ is a finite regular Borel measure and $\bigcap_1^\infty (M_n\setminus\{\lambda_0\}) = \emptyset$. Thus $x = 0$, and the proof is complete.

We shall use the above lemma in the form obtained in Theorem 5.2 and its corollary.

5.2. Theorem. _Let_ T _be a normal operator on a Hilbert space_ H, _and let_ $E(\cdot)$ _be the spectral measure of_ T. _If_ F _is a closed subset of the complex plane, then_

$$E(F)H = \cap\{(\lambda I - T)H : \lambda \in \mathbf{C}\setminus F\}.$$

Proof. Let $Z = \cap\{(\lambda I - T)H : \lambda \in \mathbf{C}\setminus F\}$. Because the spectrum of T restricted to $E(F)H$ is contained in F it follows that $E(F)H = (\lambda I - T)E(F)H \subseteq (\lambda I - T)H$ for all λ not in F. The result will now follow if we prove that $(I - E(F))Z = \{0\}$. Let (U_n) be an increasing sequence of closed subsets of $\mathbf{C}\setminus F$ such that $\bigcup_1^\infty U_n = \mathbf{C}\setminus F$. If $E(U_n)Z = \{0\}$ for all n, then $E(\mathbf{C}\setminus F)Z = (I - E(F))Z = \{0\}$ since $E(\cdot)$ is a Borel measure. Now

$$E(U_n)Z = E(U_n)[\cap\{(\lambda I - T)H : \lambda \in \mathbf{C}\setminus F\}]$$
$$\subseteq \cap\{E(U_n)(\lambda I - T)H : \lambda \in \mathbf{C}\setminus F\}$$
$$\subseteq \cap\{(\lambda I - T)E(U_n)H : \lambda \in U_n\}.$$

The spectrum of the restriction of T to $E(U_n)H$ is contained in U_n, so Lemma 5.1 implies that the last intersection is zero. This proves the theorem.

5.3. **Corollary.** Let T and R be normal operators on Hilbert spaces X and Y, respectively, and let S be a linear operator from X into Y with $ST = RS$. Then $SE_T(\mathbf{C} \setminus U)X \subseteq E_R(\mathbf{C} \setminus U)Y$ and $E_R(U)SE_T(U) = E_R(U)S$ for all open subsets U of $\mathbf{C}$, where $E_T(\cdot)$ and $E_R(\cdot)$ are the spectral measures of T and R, respectively.

Proof. By Theorem 5.2 we have

$$S(E_T(\mathbf{C} \setminus U)X) = S(\cap \{(\lambda I - T)X : \lambda \in U\}) \subseteq \cap \{(\lambda I - R)Y : \lambda \in U\}$$
$$= E_R(\mathbf{C} \setminus U)Y$$

because $S(\lambda I - T) = \lambda I - R)S$. This implies that $E_R(U)SE_T(\mathbf{C} \setminus U) = 0$ and so $E_R(U)SE_T(U) = E_R(U)S$.

5.4. **Remark.** Results like 5.2 and 5.3 have been proved for generalized scalar operators (see Colojoară and Foias [26]) by Vrbová [129]. We break up an operator intertwining with a pair of normal operators into a sum of a continuous intertwining operator and discontinuous intertwining operators associated with critical eigenvalues. A natural question to ask is whether this result extends to more general cases when Y has $\{0\}$ as the only R-divisible subspace.

5.5. **Theorem.** Let T and R be normal operators on Hilbert spaces X and Y, respectively, let S be a linear operator from X into Y satisfying $ST = RS$. Then $S = B + K_1 + \ldots + K_n$, where B is a continuous linear operator from X into Y and $\lambda_1, \ldots, \lambda_n$ are critical eigenvalues of (T, R) such that for $j = 1, \ldots, n$ there is a discontinuous linear operator K_j from X into Y with $K_j(T - \lambda_j I) = 0 = (R - \lambda_j I)K_j$.

Proof. For each open subset U of $\mathbf{C}$ we let $X(U) = E_T(\mathbf{C} \setminus U)X$ and $Y(U) = E_R(\mathbf{C} \setminus U)Y$, where the notation of 5.3 applies. The spaces $X(U)$ and $Y(U)$ satisfy the conditions 2.2 so that by Theorem 2.3 the operator S has at most a finite number of discontinuity points in $\mathbf{C}$. Because $Y = E_R(U)Y \oplus E_R(\mathbf{C} \setminus U)Y$, the quotient space $Y/E_R(\mathbf{C} \setminus U)Y$ is naturally isomorphic to $E_R(U)Y$ for all open subsets U of $\mathbf{C}$. Thus λ is a discontinuity point of S if and only if $E_R(U)S$ is discontinuous for all open subsets U of $\mathbf{C}$ containing λ.

Let $F = \{\lambda_1, \ldots, \lambda_n\}$ be the finite set of discontinuity points of S. Let $B = E_R(\mathbf{C}\backslash F)S$, and $K_j = E_R(\{\lambda_j\})S$ for $j = 1, \ldots, n$. Clearly $S = B + K_1 + \ldots + K_n$. We prove the remaining properties of B and the K_j. Let

$$\mathcal{G} = \{V : V \text{ is an open subset of } \mathbf{C}, \text{ and } E_R(V)S \text{ is continuous}\}.$$

Then $\cup \mathcal{G} = \mathbf{C}\backslash F$ by the definition of F. Each compact subset L of $\mathbf{C}\backslash F$ is covered by a finite number $V_1, \ldots, V_n$ of elements of $\mathcal{G}$, and hence

$$\begin{aligned} E_R(L)\mathfrak{S} &\subseteq E_R(V_1 \cup \ldots \cup V_n)\mathfrak{S} \\ &\subseteq E_R(V_1)\mathfrak{S} + \ldots + E_R(V_n)\mathfrak{S} = \{0\} \end{aligned}$$

because $E_R(V)\mathfrak{S} = \{0\}$ for all V in $\mathcal{G}$. Since $E_R(\cdot)$ is a bounded regular Borel measure $E_R(\mathbf{C}\backslash F)x$ is the limit of $E(L)x$ over the net of all compact subsets L of $\mathbf{C}\backslash F$, it follows that $E_R(\mathbf{C}\backslash F)\mathfrak{S} = \{0\}$. This shows that B is continuous. Further $BT = E_R(\mathbf{C}\backslash F)ST = E_R(\mathbf{C}\backslash F)RS = RE_R(\mathbf{C}\backslash F)S = RB$. Similarly $K_jT = TK_j$ for $j=1, \ldots, n$.

Let U_j be an open neighbourhood of λ_j containing no other points of F. Then $E_R(U_j)S = E_R(U_j)B + K_j$ because $E_R(U_j)E_R(\{\lambda_k\})$ is equal to $E_R(\{\lambda_j\})$ if $k = j$ and is zero if $k \neq j$. Since $E_R(U_j)S$ is discontinuous and $E_R(U_j)B$ is continuous, K_j is discontinuous. We have $K_jX \subseteq \operatorname{Ker}(R - \lambda_jI)$ since $(R - \lambda_jI)E_R(\{\lambda_j\}) = 0$. Therefore λ_j is an eigenvalue of R because K_j is non-zero. From the equation $E_R(\{\lambda_j\})(R - \lambda_jI) = 0$ we obtain $K_j(T - \lambda_jI)X = E_R(\{\lambda_j\})S(T - \lambda_jI)X \subseteq E_R(\{\lambda_j\})(R - \lambda_jI)Y = \{0\}$. If $(T - \lambda_jI)X$ were of finite codimension in X, then it would be closed in X by Lemma 3.3; hence K_j would be continuous on X contradicting the discontinuity of K_j. Thus each λ_j is a critical eigenvalue of (T, R), and the proof is complete.

5.7. **Corollary.** Let T and R be normal operators on Hilbert spaces X and Y, respectively. Then there is a discontinuous linear operator S from X into Y satisfying $ST = RS$ if and only if (T, R) has a critical eigenvalue.

5.8. **Remarks.** The results in this section are taken from Johnson [58], and Pták and Vrbová [100]. Johnson proves Theorem 2.3 by a direct rolling hump argument based on critical eigenvalues and a theorem like Theorem 5.2. Theorem 5.2 is taken from Pták and Vrbová [100]. Similar methods have been used for operators (T, R) which have a large lattice of closed invariant subspaces closely related to the spectral properties of T and R [58], [69], [128].

3 · Homomorphisms

The main research in automatic continuity has been on the continuity properties of a homomorphism from one Banach algebra into another. Some of the results on the automatic continuity of homomorphisms between Banach algebras are given in this chapter. We begin the chapter with Johnson's theorem on the uniqueness of the complete norm topology on a Banach space that is an irreducible module over a Banach algebra such that algebra multiplication on the module is continuous [59] (§6). From this the continuity of isomorphisms between semisimple Banach algebras follows easily [59]. In Section 7 we prove a result of Kaplansky [74] on the decomposition of a ring (i. e. additive) isomorphism between two semisimple Banach algebras using automatic continuity methods. Both Sections 6 and 7 depend on Theorem 2. 3. Section 8 contains a brief discussion of the relationship between discontinuous derivations and discontinuous automorphisms, and a proof of the existence of discontinuous derivations from the disc algebra into a Banach module over it [28].

Sections 9 to 12 are concerned with homomorphisms from C*-algebras and, in particular, from the Banach algebra of continuous complex valued functions on a compact Hausdorff space. The main technical result is Theorem 9. 3 (Bade and Curtis [7]) on which Sections 10 and 12 are based. Section 10 is devoted to proving the important theorem of Bade and Curtis [7] on the decomposition of homomorphisms from $\mathbf{C}(\Omega)$ into a Banach algebra into continuous and highly discontinuous parts. Lemma 1. 6 is applied in Section 11 to a homomorphism from $\mathbf{C}_0(\Psi)$ into a radical Banach algebra to prove that the image of $\mathbf{C}_0(\Psi)$ contains no nilpotent elements, where Ψ is a locally compact Hausdorff space. In Section 12 we use Theorem 9. 3 to obtain several results on the automatic continuity of homomorphisms from C*-algebras.

6. Isomorphisms of semisimple Banach algebras

The main result of this section is the uniqueness of the norm topology on a Banach space that is an irreducible module over a Banach algebra such that algebra multiplication on the module is continuous (Johnson [59]) (Theorem 6.9). We begin by developing a little of the theory of irreducible modules over a Banach algebra. For a full account see Bonsall and Duncan [16] or Rickart [103]. We end the section with a discussion of the properties of the spectrum of elements in the separating space and apply it to homomorphisms into strongly semisimple Banach algebras.

Throughout this section A is a Banach algebra over the complex field, though the representation theory can be developed in greater generality [18], [51], [103].

6.1. Definitions. A linear space X is defined to be a <u>(left) A-module</u> [right A-module] if there is a map $A \times X \to X : (a, x) \mapsto a.x$ $[(a, x) \mapsto x.a]$ such that $x \mapsto a.x$ $[x \mapsto x.a]$ is a linear operator on X for each a in A, and the map from a to this linear operator is a homomorphism [antihomomorphism] from A into the algebra of linear operators on X. An A-<u>bimodule</u> is a left A-module X and right A-module for which $a(xb) = (ax)b$ for all a, b in A and x in X.

A <u>normed</u> A-<u>module</u> (<u>Banach</u> A-<u>module</u>) X is a normed space (Banach space) X that is an A-module for which there is a constant N such that $\|ax\| \leq N\|a\|\,\|x\|$ for all a in A and x in X. If X is a normed A-module, then the map θ defined by $\theta(a)(x) = a.x$ maps A into the algebra of continuous linear operators on X and θ is continuous. Conversely using a continuous homomorphism θ from a Banach algebra A into $\mathcal{L}(X)$, where X is a normed space, we may define X to be a normed A-module by $a.x = \theta(a)(x)$. A Banach A-bimodule X is a Banach space and an A-bimodule so that there exists a constant N satisfying $\|ax\| \leq N\|a\|.\|x\|$ and $\|xa\| \leq N\|a\|.\|x\|$ for all a in A and x in X.

An A-module X is <u>(algebraically) irreducible</u> if $AX \neq \{0\}$ and if $\{0\}$ and X are the only A-submodules of X. A left ideal J in A is modular if there is an e in A so that $xe - x$ is in J for all x in A, and e is called a <u>right modular identity</u> for J. We shall refer to maximal proper ideals as maximal ideals.

Note that an A-module is irreducible if and only if for each non-zero x in X we have $A.x = X$.

6.2. Lemma. (a) *If J is a maximal modular left ideal in A, then with the multiplication $a.(b + J) = ab + J$ for all a, b in A the linear space A/J is an irreducible A-module.*

(b) *If X is an irreducible A-module, then for each non-zero x in X the set $J = \{a \in A : ax = 0\}$ is a maximal modular left ideal in A.*

Proof. (a) Since $a(x + J) = ax + J$ for all a and x in A, there is a one-to-one correspondence between left ideals K containing J and submodules of A/J given by $K \longleftrightarrow K/J$. Thus there are no proper submodules of A/J, i.e., no submodules of A/J other than A/J and the zero module. Let e be a right modular identity for J. Then $xe - x$ is in J for each x in A so there is an x in A with xe not in J. Thus $A.(A/J) \neq \{J\}$, and A/J is an irreducible A-module.

(b) Because X is an A-module, J is a left ideal in A. The irreducibility of X implies that there is an e in A so that $ex = x$. Hence $(ae - a)x = ax - ax = 0$, and $ae - a$ is in J for all a in A. Therefore J is a modular left ideal not equal to A. If K is a left ideal in A with $A \supseteq K \supset J$, then $K.x$ is a non-zero A-submodule of X. Hence $K.x = X$. If a is in A, then there is a b in K so that $bx = ax$. Thus $a - b$ is in J and so $a = (a - b) + b$ is in K. Therefore J is a maximal left ideal.

6.3. Lemma. *A maximal modular left ideal in a Banach algebra is closed.*

Proof. Let J be a maximal modular left ideal in a Banach algebra A, and let e be a modular right identity for J. Suppose that x in J satisfies $\|e - x\| < 1$. Let $u = \sum_1^\infty (e - x)^n$. Then $u - u(e - x) = e - x$ so that $e = x + ux + u - ue$ is in J. Hence J is equal to A which is a contradiction. Thus $J \cap \{x \in A : \|e - x\| < 1\}$ is empty so J^- is a proper closed modular left ideal in A containing J. The maximality of

J implies that $J = J^-$. This proves the lemma.

6.4. Remark. (a) The above lemma is the vital link between the algebraic and topological properties which lead to the uniqueness of the complete norm topology on a semisimple Banach algebra.

(b) In an algebra with identity every ideal is modular. If K is a proper modular left ideal in A with right modular identity e, then an application of Zorn's lemma gives a maximal left ideal J in A containing K and not containing e. Thus maximal modular left ideals exist in an algebra with identity.

6.5. Lemma. Let X be an irreducible A-module. If $\mathbf{D}$ is the set of all linear operators T on X such that $T(ax) = aT(x)$ for all a in A and x in X, then $\mathbf{D} = \mathbf{C}I$, where I is the identity operator.

Proof. We begin by proving that $\mathbf{D}$ is a division algebra. Clearly $\mathbf{D}$ is an algebra of not necessarily continuous linear operators. Let T be a non-zero element in $\mathbf{D}$. Since TX and $\operatorname{Ker} T$ are submodules of X and $TX \neq \{0\}$, it follows that $TX = X$, and $\operatorname{Ker} T = \{0\}$. Hence T^{-1} exists as a linear operator on X. Because

$$T^{-1}(ax) = T^{-1}(aTT^{-1}x) = T^{-1}T(aT^{-1}x) = aT^{-1}(x)$$

for all a and x, the operator T^{-1} is in $\mathbf{D}$. We now define a norm on $\mathbf{D}$ to make it into a normed algebra. Let z be a non-zero element in X, and let $J = \{a \in A : az = 0\}$. Then the map $A/J \to X : a + J \mapsto az$ is an isomorphism of A-modules. We obtain a norm on X by defining $\|x\| = \|a + J\| = \inf\{\|b\| : b \in A,\ bz = x\}$ where $az = x$. Let T be in $\mathbf{D}$, and choose an f in A with $fz = Tz$. If $bz = x$ with b in A, then $Tx = Tbz = bTz = bfz$, and $\|bf\| \leq \|b\| . \|f\|$. Taking the infimum over all such b we obtain $\|Tx\| \leq \|f\| . \|x\|$ for all x in X. Hence $\mathbf{D}$ is an algebra of continuous linear operators on X with the norm defined above. The Gelfand-Mazur Theorem implies that $\mathbf{D}$ is isomorphic to $\mathbf{C}$ under the natural isomorphism from $\mathbf{C}$ into $\mathbf{D}$ so $\mathbf{D} = \mathbf{C}I$. This completes the proof.

If the Banach algebra were just a real Banach algebra, then $\mathbf{D}$

would be isomorphic to the reals, complexes, or quaternions [18], [103].

6.6. **Lemma.** Let X be an irreducible A-module. If $x_1, \ldots, x_n$ are linearly independent in X, then there is an a in A such that $ax_1 = \ldots = ax_{n-1} = 0$, and $ax_n \neq 0$ $(n = 2, 3, \ldots)$.

Proof. We prove the lemma by induction on n over all Banach algebras, all irreducible modules over them, and all linearly independent subsets of n elements of the modules. We start with $n = 2$. Suppose that a in A with $ax_1 = 0$ implies that $ax_2 = 0$. Since $Ax_1 = X$, we can define $T : X \to X$ by $Ty = ax_2$ if $ax_1 = y$. Then T is a well defined linear operator from X into X. If y is in X and b is in A, then $T(by) = bax_2$, where $ax_1 = y$ because $bax_1 = by$. Thus $T(by) = b(Ty)$. Hence T is in $\mathbf{D}$, and so $T = \lambda I$ for some λ in $\mathbf{C}$ by 6.5. Therefore $a(\lambda x_1 - x_2) = 0$ for all a in A. This implies that $\lambda x_1 - x_2 = 0$ since $\{z \in X : Az = \{0\}\}$ is an A-submodule of X not equal to X (and so is equal to $\{0\}$). This contradicts the linear independence of x_1, x_2.

Suppose the inductive hypothesis holds for $n - 1$ (≥ 2). Let $x_1, \ldots, x_n$ be linearly independent in X. Let $J = \{a \in A : ax_1 = \ldots = ax_{n-2} = 0\}$, let $M = \mathbf{C}x_1 + \ldots + \mathbf{C}x_{n-2}$, and let $Y = X/M$. Then J is a left ideal in A, and Y may be regarded as a J-module by defining $a(x + M) = ax + M$ for all a in J and x in X. By the inductive hypothesis Jx is a non-zero A-submodule of X for each x not in M. Thus $Jx = X$ for each x not in M, so that Y is an irreducible J-module. Further J is a Banach algebra as it is an intersection of closed maximal modular left ideals (6.2, 6.3). By the case $n = 2$, there is a' in J with $a'x_{n-1}$ in M and $a'x_n$ not in M. Again by $n = 2$ there is b in A such that $ba'x_{n-1} = 0$ and $ba'x_n \neq 0$. Letting $a = ba'$ completes the induction, and the proof of the lemma.

6.7. **Theorem.** Let X be an irreducible module over a Banach algebra A. If $x_1, \ldots, x_n$ are linearly independent in X, and $y_1, \ldots, y_n$ are in X, then there is an a in A such that $ax_j = y_j$ for $j = 1, \ldots, n$. If A is unital, if $1.x = x$ for all x in X, and if $y_1, \ldots, y_n$ are linearly independent, then a can be chosen to be invertible.

Proof. For each j there is a b_j so that $b_j x_k = 0$ if $k \neq j$ and $b_j x_j \neq 0$ by Lemma 6.6. Because X is an irreducible A-module there is a c_j in A such that $c_j b_j x_j = y_j$. Let $a = \sum_1^n c_j b_j$. This completes the first part of the proof.

Choose and fix a non-zero element z in X, and define $\|x\| = \inf\{\|b\| : b \in A,\ bz = x\}$ for all x in X. Then as in the proof of 6.5, X with $\|\cdot\|$ is a Banach space. If a is in A and x in X, then $bz = x$ implies that $(ab)z = ax$ and $\|ab\| \leq \|a\| . \|b\|$. Taking the infimum over all such b we obtain $\|ax\| \leq \|a\| . \|x\|$. Hence X is a Banach A-module. Now suppose $y_1, \ldots, y_n$ are linearly independent. Let M be the span of $x_1, \ldots, x_n, y_1, \ldots, y_n$. Then there is an invertible linear operator T on M such that $Tx_j = y_j$ for each j. Because the spectrum of T is finite and does not contain 0 there is a branch of the logarithm which is analytic in a neighbourhood of $\sigma(T)$. Taking this logarithm of T we obtain a linear operator R on M such that $T = \exp R$. There is a b in A so that $bx = Rx$ for all x in M. The element $a = \exp b$ has the properties required in the conclusion.

The second part of the above theorem is well known for C*-algebras (see Dixmier [34, 2.8.3(ii), p. 45]).

6.8. Lemma. Let X be a normed space, let z be a non-zero element of X, and let $Y = \{T \in \mathcal{L}(X) : Tz = 0\}$. Then the map $\psi : \mathcal{L}(X)/Y \to X : T + Y \to Tz$ is a topological isomorphism.

Proof. Clearly the map ψ is continuous with $\|\psi\| \leq \|z\|$, and is one-to-one. We now prove that ψ is onto and ψ^{-1} is continuous. Let f be in X^* such that $\|f\| = 1$ and $f(z) = \|z\|$ (Hahn-Banach Theorem). If x is in X, then T defined by $T(w) = f(w)x/\|z\|$ satisfies $\|T\| = \|x\|/\|z\|$ and $Tz = x$. Hence ψ is invertible and

$$\|\psi^{-1}\| = \|z\|^{-1}.$$

6.9. Theorem. Let A be a Banach algebra, and let X be an irreducible A-module and a normed space. If $X \to X : x \mapsto ax$ is continuous for each a in A, then X is a normed-A-module, that is, there exists a real number M such that $\|ax\| \leq M \|a\| . \|x\|$ for all a in A and x in X.

Proof. Let $\theta : A \rightarrow \mathcal{L}(X)$ be defined by $\theta(a)x = ax$ for all a in A and x in X. Our hypotheses imply that θ is an (algebraic) homomorphism from A into $\mathcal{L}(X)$. If X is finite dimensional, then Ker θ is of finite codimension in A. Also Ker θ is an intersection of maximal modular left ideals so Ker θ is closed by Lemma 6.3. Hence θ is continuous.

This leaves the important case when X is infinite dimensional. We first prove that $A \rightarrow X : a \rightarrow ax$ is continuous for each x in X. Suppose that for some z in X the map $a \mapsto az$ is discontinuous. If x is non-zero and $a \mapsto ax$ is continuous, then there is a b in A with $bx = z$ so that $az = abx$, and $a \mapsto az$ is continuous; it is the product of the two continuous linear operators $a \mapsto ab$ and $c \mapsto cx$. Thus $a \mapsto ax$ is discontinuous for all non-zero x in X.

We now choose a linearly independent sequence (x_n) from X. We shall apply Theorem 2.3 with $X = A$, $Y = \mathcal{L}(X)$, Ω the set of positive integers with the discrete topology, and Γ the set of singleton subsets of Ω. If $\{j\}$ is a singleton subset of the positive integers, we let

$$A(\{j\}) = \{a \in A : ax_j = 0\},$$

and

$$Y(\{j\}) = \{T \in \mathcal{L}(X) : Tx_j = 0\}.$$

Then $A(\{j\})$ is closed in A, because it is a maximal modular left ideal (6.2 and 6.3). Also $Y(\{j\})$ is closed in $Y = \mathcal{L}(X)$. If $F_1, \ldots, F_n$ are pairwise disjoint singleton subsets of Ω, then

$$A(F_1) \cap \ldots \cap A(F_{n-1}) + A(F_n) = A$$

because the left hand side is a modular left ideal in A that properly contains a maximal modular left ideal (6.7). By the construction of A(F), it follows that $\theta A(F)$ is contained in Y(F). Theorem 2.3 now shows that there are only a finite number of discontinuity points. If j is not a discontinuity point of θ in Ω, then the map

$$A \to \mathcal{L}(X)/Y(\{j\}) : a \mapsto \theta(a) + Y(\{j\})$$

is continuous. By Lemma 6.8 the map $A \to X : a \mapsto \theta(a)x_j = ax_j$ is continuous. This gives a contradiction.

Hence $A \to X : a \mapsto \theta(a)x$ is continuous for each x in X, and $\theta(a)$ is a continuous linear operator on X for each a in A. The uniform boundedness theorem implies that θ is continuous, and the proof is complete.

6.10. Corollary. An irreducible module over a Banach algebra has a unique Banach space topology such that multiplication by each element in the algebra is a continuous linear operator on the module.

Proof. Let A be a Banach algebra, let X be an irreducible A-module, and let z be a non-zero element in X. Then the map $A/J \to X : a + J \mapsto az$, where $J = \{a \in A : az = 0\}$, is an A-module isomorphism from A/J onto X. Because J is a maximal modular left ideal in A (Lemma 6.2(b)), J is closed (Lemma 6.3). The quotient Banach space norm on A/J is a Banach A-module norm, and may be carried to X by the A-module isomorphism. Hence X is a Banach A-module with norm $\|x\| = \inf\{\|b\| : b \in A, bz = x\}$.

If $|\cdot|$ is a complete norm on X such that $x \mapsto ax$ is continuous for each a in A, then there is a constant M so that $|ax| \leq M\|a\|\,|x|$ for all a in A and x in X (Theorem 6.9). Therefore $|x| \leq M\|b\|\,|z|$ for all b in A with $bz = x$, so $|x| \leq M\|x\| . |z|$ for all x in X. This inequality holds without the hypothesis that $|.|$ is complete. When $|.|$ is complete an application of the open mapping theorem gives an m such that $\|x\| \leq m|x|$ and completes the proof.

6.11. Definition. The (Jacobson) radical of a Banach algebra is the algebra if there are no maximal modular left ideals and is the intersection of all maximal modular left ideals in the algebra if there are such ideals, and the algebra is said to be semisimple if the radical is $\{0\}$. If there are no maximal modular left ideals in the algebra, then the algebra is said to be a <u>radical</u> algebra.

The radical is clearly a closed left ideal by Lemma 6.3, and is actually a two sided ideal. For let a be in the radical, and b be in the algebra. If X is an irreducible A-module, then $aX = \{0\}$ by Lemma 6.2(b) so that $(ab)X = a(bX) \subseteq aX = \{0\}$. Thus ab is in the radical by 6.2(a). An ideal P in an algebra A is said to be <u>(left) primitive</u> if there is a maximal modular left ideal J in A such that $P = \{a \in A : aA \subseteq J\}$.

6.12. Corollary. <u>An epimorphism from a Banach algebra to a semisimple Banach algebra is continuous.</u>

Proof. Let θ be an epimorphism from a Banach algebra A to a semisimple Banach algebra B, and let J be a maximal modular left ideal in B. Then we may regard B/J as an A-module by $a.(b + J) = \theta(a)b + J$ for all a in A and b in B. Now B/J is a Banach space, the module operations $b + J \mapsto a.(b + J)$ are continuous for each a in A, and B/J is an irreducible A-module because θ is an epimorphism. By Theorem 6.9 the map $A \to B/J : a \mapsto \theta(a)e+J = \theta(a)+J$ is continuous, where e is a right modular identity for J in B. Thus $\mathfrak{S}(\theta)$ is contained in J, so that $\mathfrak{S}(\theta) = \{0\}$. This implies that θ is continuous, and the proof is complete.

An <u>involution</u> * on a Banach algebra A is a conjugate linear operator $a \mapsto a^*$ on A such that $(ab)^* = b^*a^*$ for all a and b in A, and an <u>automorphism</u> θ on A is an algebraic isomorphism from A onto A.

6.13. Corollary. <u>A semisimple Banach algebra has a unique complete algebra-norm topology, and any involutions or automorphisms on the algebra are continuous.</u>

Proof. Let $\|\cdot\|$ and $|\cdot|$ be complete norms on a semisimple Banach algebra A. Apply Corollary 6.12 to the identity operator from $(A, \|\cdot\|)$ to $(A, |\cdot|)$ and to the inverse of this operator. Corollary 6.12 deals with automorphisms. If * is an involution on A, let

$|x|_1 = \|x^*\|$. Then $|\cdot|_1$ is a complete algebra norm on A, so is equivalent to $\|\cdot\|$, and hence $*$ is continuous.

6.14. Remarks. The previous results on the continuity of homomorphisms were obtained by studying the global properties of θ, and their relation to irreducible representations of the algebra. Different results may be obtained by considering the detailed structure of the spectra of individual elements in the separating space; these methods are local. If A is commutative and θ is a homomorphism from A onto a dense subalgebra of a Banach algebra B, then $\mathfrak{S}$ is contained in the radical of B so each element of $\mathfrak{S}$ has spectrum $\{0\}$. The best general result I know of is Theorem 6.16 (see Rickart [103] and Barnes [12]). In the proof we shall require the following lemma due to Newburgh [95].

6.15. Lemma. *Let A be a unital Banach algebra, let x be in A, and let (x_n) be a sequence in A converging to x. If V is a neighbourhood of 0 in $\mathbf{C}$, and W is a non-empty open and closed subset of $\sigma(x)$, then there are positive integers N_1 and N_2 such that $\sigma(x_n) \subseteq V + \sigma(x)$ for all $n \geq N_1$, and $\sigma(x_n) \cap (W + V)$ is non-empty for all $n \geq N_2$.*

Proof. We may assume V is open by replacing it by its interior. Suppose there is no integer N_1. Then there is a subsequence (z_n) of (x_n), and a sequence (λ_n) of complex numbers such that λ_n is in $\sigma(z_n)\backslash(\sigma(x) + V)$. The sequence (λ_n) is bounded because $|\lambda_n| \leq \|z_n\|$ for all n. Hence there is a subsequence of (λ_n) that converges to a complex number λ. We denote this subsequence by (λ_n). Then λ is not in $\sigma(x) + V$, and so is not in $\sigma(x)$. Since

$$\|(\lambda 1 - x) - (\lambda_n 1 - z_n)\| \leq |\lambda_n - \lambda| + \|x - z_n\|$$

tends to zero and $\lambda 1 - x$ is invertible, it follows that $\lambda_n 1 - z_n$ is invertible for all sufficiently large n. This is a contradiction.

Suppose there is no N_2. Then there is a subsequence (y_n) of (x_n) such that $\sigma(y_n) \cap (W + V) = \emptyset$ for all n. Let γ be a closed path enclosing W, contained in $W + V$, enclosing no points of $\sigma(x)\backslash W$, and disjoint from $\sigma(x)$. The inequality

$$\|(\lambda 1 - x)^{-1} - (\lambda 1 - y_n)^{-1}\|$$
$$\leq \{1 - \|x - y_n\| \|(\lambda 1 - x)^{-1}\|\}^{-1} \|(\lambda 1 - x)^{-1}\|^2 \|x - y_n\|$$

for λ in γ and the boundedness of the set $\{\|(\lambda 1 - x)^{-1}\| : \lambda \in \gamma\}$ imply that $(\lambda 1 - y_n)^{-1}$ converges to $(\lambda 1 - x)^{-1}$ uniformly in γ. Hence $\int_\gamma (\lambda 1 - x)^{-1} d\lambda = \lim \int_\gamma (\lambda 1 - y_n)^{-1} d\lambda$. Since γ encircles no elements in $\sigma(y_n)$, $\int_\gamma (\lambda 1 - y_n)^{-1} d\lambda$ is zero for all n. Since γ encloses a non-empty open and closed subset of $\sigma(x)$, $\int_\gamma (\lambda 1 - x)^{-1} d\lambda$ is a non-zero element in A. This contradiction completes the proof.

Recall that z is a joint topological zero divisor in A if there is a sequence (w_n) in A with $\|w_n\| = 1$ for all n and $\|w_n z\| + \|z w_n\| \to 0$ as $n \to \infty$. Further the boundary of the set of invertible elements in a unital Banach algebra is contained in the set of joint topological zero divisors [18], [103].

6.16. Theorem. Let θ be a homomorphism from a Banach algebra A into a Banach algebra B, and let x be in the separating space $\mathfrak{S}$ of θ. Then the spectrum of x is a connected subset of **C** containing 0, and $\lambda 1 - x$ is a joint topological divisor of zero in B for each λ in $\sigma(x)$.

Proof. We may assume that A and B are unital by adjoining identities if necessary. Suppose 0 is not in $\sigma(x)$. Let (a_n) be a sequence in A with $\|a_n\| \to 0$ and $\|\theta(a_n) - x\| \to 0$. Then choose a compact neighbourhood V of 0 in **C** so that 0 is not in $\sigma(x) + V$. For n sufficiently large $\sigma(a_n) \cap (\sigma(x) + V)$ is empty, and so $\sigma(\theta(a_n)) \cap (\sigma(x)+V)$ is empty because $\sigma(\theta(a_n)) \subseteq \sigma(a_n)$. This contradicts Lemma 6.15.

If $\sigma(x)$ is not connected, we let W be an open and closed subset of $\sigma(x)$ with 0 not in W, and choose a compact neighbourhood V of 0 so that 0 is not in $W + V$. For n sufficiently large $\sigma(\theta(a_n)) \cap (W + V)$ is empty as before, and again there is a contradiction.

Suppose λ is non-zero and is in $\sigma(x)$. Then $\lambda 1 - x$ is the limit of the sequence $(\lambda 1 - \theta(a_n))$, whose elements are invertible for sufficiently large n. Hence $\lambda 1 - x$ is in the boundary ∂G of the set of invertible elements in B. Therefore $\lambda 1 - x$ is a joint topological zero divisor. If $\sigma(x)$ contains more than one point, then 0 is in $(\sigma(x)\setminus\{0\})^-$ so x is in $(\partial G)^- = \partial G$. If $\sigma(x) = \{0\}$, then x is in ∂G because $\lambda 1 - x$ is invertible for all non-zero λ. In both cases x is a joint topological zero divisor, and the proof is complete.

6.17. Definition. The intersection of the maximal modular two sided ideals in a Banach algebra B is called the strong radical of B. If the strong radical of B is zero, then B is called strongly semisimple.

One can prove that a maximal modular two sided ideal in a Banach algebra is closed (see Lemma 6.3, and [103]).

6.18. Theorem. Let A be a unital Banach algebra, and let θ be a homomorphism from A onto a dense subalgebra of a strongly semisimple Banach algebra B. Then θ is continuous.

Proof. Because A has an identity, B has an identity. Let J be a maximal two sided ideal in B, and let Q be the natural quotient map from B onto B/J. Then $Q\theta$ is a homomorphism from A into B/J, and $\mathfrak{S}(Q\theta) = (Q\,\mathfrak{S}(\theta))^-$ is a (closed) two sided ideal in B/J. If $\mathfrak{S}(Q\theta)$ is non-zero, then $\mathfrak{S}(Q\theta)$ contains the identity of B/J. This contradicts Theorem 6.16. Hence $\mathfrak{S}(\theta) \subseteq J$, and so $\mathfrak{S}(\theta) = \{0\}$ and thus θ is continuous.

6.19. Remarks. (a) The hypothesis of strong semisimplicity in Theorem 6.18 may be weakened to semisimplicity provided an additional assumption is added to give a contradiction when 6.16 is applied in the proof (see Barnes [12]).

(b) Theorem 6.18 and Corollary 6.12 suggest the following question. Is a homomorphism from a Banach algebra onto a dense subalgebra of a semisimple Banach algebra continuous?

(c) The results in this section are taken from Johnson [59], Rickart [101], [103], Yood [132], Barnes [12], and Newburgh [95].

7. Ring isomorphisms of semisimple Banach algebras

7.1. Remarks. Throughout these notes we restrict attention to complex linear operators except in this section, where we prove a result of Kaplansky [74, p. 380] on the decomposition of a ring isomorphism between semisimple Banach algebras. Our method is a combination of those of Kaplansky [74] and Johnson [59], and is similar to the proof of the corresponding ring derivation theorem [68]. Continuity and boundedness are equivalent for a rational linear operator, and imply that the operator is real linear. We shall apply Theorems 2.3 and 6.9 (for infinite dimensional modules) to rational linear operators; the proofs of the rational linear forms of these theorems are the same as those given. The property on which the results in this section are based is that a maximal modular left ring ideal in an algebra A is a linear subspace, and so an algebra ideal.

7.2. Lemma. Maximal modular left ring ideals and (left) primitive ring ideals in an algebra are linear subspaces.

Proof. Let L be a maximal modular left ring ideal in an algebra A. Then L is clearly contained in the left algebra ideal $\{x \in A : Ax \subseteq L\}$, and this algebra ideal is not equal to A since it does not contain a modular right identity for L. By the maximality of L as a left ring ideal, L is equal to $\{x \in A : Ax \subseteq L\}$ so L is a linear subspace of A.

If $P = \{a \in A : aA \subseteq L\}$, if a is in P, and if λ is a scalar, then $(\lambda a)A = a(\lambda A) \subseteq aA \subseteq L$. Hence P is a linear subspace of A.

7.3. **Lemma.** Let A and B be Banach algebras, let θ be a ring epimorphism from A to B, let P be a primitive ideal in B, and let $Q = Q_P : B \to B/P$ be defined by $Q(b) = b + P$. If P is of infinite codimension in B, then $Q\theta$ is a continuous real linear operator.

Proof. Let M be a maximal modular left ideal in B such that $P = \{b \in B : bB \subseteq M\}$. Then M is of infinite codimension in B because P is of infinite codimension in B. Regard B/M as a left-A-module by defining $a.(b + M) = \theta(a)b + M$ for all a in A and b in B. The central portion of the proof of Theorem 6.9 holds for rational linear maps so that $A \to B/M : a \mapsto \theta(a)b + M$ is continuous for all b in B. Hence $\mathfrak{S}(\theta)b + M = M$ for all b in B, so that $\mathfrak{S}(\theta)$ is contained in P. As the closed graph theorem is true for rational linear maps between Banach spaces, $Q\theta$ is continuous. Hence $Q\theta$ is a real linear operator, and the proof is complete.

In the next continuity lemma we require a corollary of the Wedderburn decomposition theorem for semisimple finite dimensional algebras (see [68, Lemma 3.1]).

7.4. Lemma. _If $P_1, \ldots, P_m$ are distinct primitive ideals of finite codimension in an algebra_ A _over the complex field, then_ $A/P_1 \cap \ldots \cap P_m$ _is isomorphic to_ $A/P_1 \oplus \ldots \oplus A/P_m$, _and_

$$\cap\{P_j : 1 \le j \le n\} + \cap\{P_j : n+1 \le j \le m\} = A$$

for $n = 1, 2, \ldots, m$.

Proof. By replacing A by $A/P_1 \cap \ldots \cap P_m$ we may assume that $P_1 \cap \ldots \cap P_m = \{0\}$. By the Wedderburn theory for finite dimensional semisimple algebras over the complex field, A is isomorphic to $M_1 \oplus \ldots \oplus M_k$, where the M_j are full matrix algebras over the complex field. Because P_j is a primitive ideal in $M_1 \oplus \ldots \oplus M_k$, the ideal P_j is equal to $\oplus\{M_i : i \ne h_j\}$ for some h_j, and M_{h_j} is isomorphic to A/P_j. Since the P_j are distinct we have $k = m$. Thus $M_1 \oplus \ldots \oplus M_m$ is isomorphic to $A/P_1 \oplus \ldots \oplus A/P_m$. Also

$$\cap\{P_j : 1 \le j \le n\} + \cap\{P_j : n+1 \le j \le m\} = A$$

because each M_i occurs in one and only one of these intersections. This completes the proof.

7.5. Lemma. _Let_ A _and_ B _be Banach algebras, and let_ θ _be a ring epimorphism from_ A _to_ B. _Then_ $Q_P\theta$ _is discontinuous for at most a finite number of primitive ideals_ P _that are of finite codimension in_ B.

Proof. Let Ω be the set of primitive ideals of finite codimension in B, and let Ω have the discrete topology. A primitive ideal of finite codimension is a maximal modular two sided ideal, so that $\theta^{-1}(P)$ is a maximal modular two sided ideal in A for each P in Ω. Therefore $\theta^{-1}(P)$ is closed in A for each P in Ω. If $\{P\}$ is a singleton subset of Ω, we let $A(\{P\}) = \theta^{-1}(P)$ and $B(\{P\}) = P$. Then $A(\{P\})$ and $B(\{P\})$ are closed linear subspaces of A and B, respectively. The maps $\{P\} \mapsto A(\{P\})$ and $\{P\} \mapsto B(\{P\})$ (Lemma 7.4), and the map θ satisfy the Conditions 2.2. The result then follows from Theorem 2.3.

7.6. **Theorem.** *Let A and B be semisimple Banach algebras, and let θ be a ring isomorphism from A to B. Then there are closed two sided ideals A_1, A_2, A_3 in A such that $A = A_1 \oplus A_2 \oplus A_3$, A_1 is finite dimensional, θ is complex linear on A_2, θ is conjugate linear on A_3, and θ is continuous on $A_2 \oplus A_3$.*

Proof. We may assume that A and B have identities, by adjoining identities to A and B, if necessary, and lifting θ to a homomorphism between the larger algebras. By Lemmas 7.3 and 7.5 the operators $Q_P\theta$ are continuous and real linear for all primitive ideals in B except for a finite number of primitive ideals of finite codimension. Let $\{P_1, \ldots, P_n\}$ be this finite exceptional set of primitive ideals in B, and let $R_j = \theta^{-1}(P_j)$ for $j = 1, \ldots, n$. We let $A_2 \oplus A_3$ be $\bigcap_1^n R_j$, and shall obtain A_1 by induction over $R_1, \ldots, R_n$. We shall then construct A_2 and A_3.

By Lemma 7.4 there is an x in A so that $1 - x$ is in R_1, and x is in R_j for $j = 2, \ldots, n$. We let π be the isomorphism from A/R_1 to B/P_1 induced by θ. Now A/R_1 and B/P_1 are isomorphic to the algebra of $m \times m$ complex matrices for some positive integer m, and so have centres isomorphic to the complex numbers. As π preserves centres it induces an automorphism ψ of the complex numbers, and ψ is discontinuous since π is discontinuous; for if ψ were one of the two continuous automorphisms on **C**, then π would be continuous as it would be real linear on a finite dimensional space. Hence there is a complex number λ so that $|\psi(\lambda)| > 2|\lambda|\,\|x\|$. Let $y = \theta(\lambda x) . \psi(\lambda)^{-1}$. By definition of ψ we have

$$\pi^{-1}(\theta(\lambda x)\psi(\lambda)^{-1} + P_1) = \pi^{-1}\{(\theta(\lambda x) + P_1)(\theta(\lambda^{-1}1) + P_1)\}$$
$$= (\lambda x + R_1)(\lambda^{-1}1 + R_1) = x + R_1 = 1 + R_1.$$

Therefore $y - 1$ is in P_1. Because x is in R_j for $j = 2, \ldots, n$ and R_j is a linear subspace of A, it follows that y is in P_j for $j=2, \ldots, n$.

Let P_0 be a primitive ideal in B not equal to any of $P_1, \ldots, P_n$ and let $R_0 = \theta^{-1}(P_0)$. Then the induced isomorphism from A/R_0 to B/P_0 is continuous and real linear, and so preserves the spectral radius. Since the spectral radius of x is bounded by $\|x\|$, the spectral radius of $y + P_0$ is bounded by $\|x\| . |\lambda| . |\psi(\lambda)^{-1}| < \frac{1}{2}$. Now the spectrum of y in B is the union of the spectra of $y + P$ in B/P as P runs over all the primitive ideals in B. With the exception of an isolated point at 1, which arises from the ideal P_1, the spectrum of y lies in the open disc of radius $\frac{1}{2}$ and centre 0. Let e be the idempotent given by the single variable analytic functional calculus corresponding to the disconnection of the spectrum of y described above so that $e\hat{}(1) = 1$ and $e\hat{}$ is zero in the disc of radius $\frac{1}{2}$, where $e\hat{}$ denotes the Gelfand transformation of e in the unital Banach algebra generated by y. Then $e - 1$ is in P_1 and e is in P for all other primitive ideals P in B. The semisimplicity of B implies that e is a central idempotent in B, and hence that $B = P_1 \oplus eB$. We may now carry e back to A by θ^{-1} and obtain $A = R_1 \oplus \theta^{-1}(e)A$. Repeating this process for $R_2, R_3, \ldots, R_n$ the decomposition $A = A_1 \oplus \bigcap_1^n R_j$ results, where A_1 is a closed finite dimensional ideal in A, and θ is continuous and real linear on $\bigcap_1^n R_j$.

By restricting attention to $\bigcap_1^n R_j$, we may assume that θ is real linear (and continuous) on A. We have to construct the final decomposition $A = A_2 \oplus A_3$. Let χ_1 be the operation of multiplication by i in A, and χ_2 be the operation of multiplication by i in B. Then $\chi_1^2 = (\theta^{-1}\chi_2\theta)^2$ is minus the identity operator on A. If χ_1 and $\theta^{-1}\chi_2\theta$ commute, the square of their product will be the identity operator. Now χ_1 and $\theta^{-1}\chi_2\theta$ each satisfy the equations $\psi(xy) = \psi(x)y = x\psi(y)$ for all x and y in A. Hence $(\chi_1\theta^{-1}\chi_2\theta)(x)y = (\theta^{-1}\chi_2\theta\chi_1)(x)y$ for all x and y in A. The semisimplicity of A ensures that $\{z \in A : zx = 0 \text{ for all } x \in A\} = \{0\}$, and hence χ_1 and $\theta^{-1}\chi_2\theta$ commute. Let $\omega = \chi_1\theta^{-1}\chi_2\theta$. Then ω is a continuous real linear operator on A, ω^2 is the identity on A, and $\omega(xy) = \omega(x)y = x\omega(y)$ for all x and y in A. We let $A_2 = (\omega - I)A$ and $A_3 = (\omega + I)A$. Then, from the properties of ω, A_2 and A_3 are closed ideals in A and $A = A_2 \oplus A_3$. We check the complex linearity of θ on A_2; the conjugate linearity of θ on A_3 is similar. It is sufficient to check multiplication by i. If x is in A, then

$$\begin{aligned}\theta(i(\omega - I)x) &= \theta(ii\theta^{-1}(i\theta(x)) - ix)\\ &= -i\theta(x) - \theta(ix) = i(i\theta(ix) - \theta(x))\\ &= i\theta((\omega - I)x).\end{aligned}$$

This completes the proof.

7.7. Corollary. Let A be a semisimple Banach algebra that does not contain a non-zero central idempotent e such that eA is finite dimensional. Then each ring isomorphism from A onto a Banach algebra is continuous.

8. Existence of discontinuous homomorphisms and derivations

We gave two methods for constructing discontinuous intertwining operators in Section 3; one by annihilating certain subspaces by an extension of linear functionals, and the other using results concerning injective modules over **C**[χ]. In this section we shall give two methods of constructing discontinuous derivations from a commutative Banach algebra into a Banach module. The first method is again an elementary annihilation technique (Lemma 8.3). The second depends on an algebraic theorem for extending derivations from a field, though Dales's original proof [28] was a direct extension argument.

We begin by giving a construction of a discontinuous monomorphism from a Banach algebra given that there is a discontinuous derivation from the algebra. This result then provides two methods of constructing discontinuous homomorphisms from the two results on discontinuous derivations that follow.

8.1. Remarks. If A is an algebra and M is an A-bimodule, then a derivation δ from A into M is a linear map from A into M satisfying $\delta(ab) = a\delta(b) + \delta(a)b$ for all a and b in A. If A is a Banach algebra and M is a Banach A-bimodule, then $|\cdot|$ defined on M by

$$|m| = \sup\{\|axb\|, \|ax\|, \|xb\|, \|x\| : a, b \in A \|a\| \leq 1, \|b\| \leq 1\}$$

is a norm on M equivalent to the given norm $\|\cdot\|$, and $|am| \leq \|a\|.|m|$ and $|ma| \leq \|a\|.|m|$ for all a in A and m in M (see Bonsall and Duncan [147, Lemma 10.3, p. 90]). Further $A \oplus M$ with norm $\|(a, m)\| = \|a\| + |m|$ and product

$$(a_1, m_1)(a_2, m_2) = (a_1a_2, a_1m_2 + m_1a_2)$$

is a Banach algebra. The proof of these two observations are routine calculations. Also $a \mapsto (a, 0)$ is an isometric monomorphism from A into $A \oplus M$, and $\{0\} \oplus M$ is a nilpotent ideal of index 2 in $A \oplus M$ (i.e. $x_1x_2 = 0$ for all x_1, x_2 in $\{0\} \oplus M$). The algebra $A \oplus M$ is commutative if and only if A is commutative and M is a commutative A-bimodule, that is, $am = ma$ for all a in A and m in M. If δ is a derivation from A into M, then $\theta : A \to A \oplus M : a \mapsto (a, \delta(a))$ is a monomorphism from A into $A \oplus M$. The algebraic ideas are well known. A direct calculation shows that $\mathfrak{S}(\theta) = \{0\} \oplus \mathfrak{S}(\delta)$. Thus we have proved the following theorem.

8.2. Theorem. Let A be a Banach algebra and let M be a Banach A-bimodule. If there is a discontinuous derivation from A into M, then there is a discontinuous homomorphism θ from A into a Banach algebra B such that $\mathfrak{S}(\theta)$ is a nilpotent ideal in B.

The following lemma is Example 1 of [7, p. 547].

8.3. Lemma. Let A be a unital Banach algebra, let θ be a character on A, and regard $\mathbf{C}$ as a Banach A-bimodule by defining $a.\lambda = \lambda.a = \theta(a)\lambda$ for all a in A and λ in $\mathbf{C}$. If the linear span Y of $(\operatorname{Ker} \theta)^2$ is not closed and of finite codimension in $\operatorname{Ker} \theta$, then there is a discontinuous derivation from A into $\mathbf{C}$.

Proof. Let δ be a discontinuous linear functional on A, chosen by Zorn's Lemma, such that δ is zero on $\mathbf{C}1 + Y$, where 1 is the identity of A. From the decomposition

$$ab = (a-\theta(a)1)(b-\theta(b)1) + \theta(a)(b-\theta(b)1) + \theta(b)(a-\theta(a)1) + \theta(ab)1$$

we obtain

$$\delta(ab) = \theta(a)\delta(b) + \theta(b)\delta(a)$$

because $a - \theta(a)1$ and $b - \theta(b)1$ are in Ker θ. Hence δ is a discontinuous derivation from A into the Banach A-module $\mathbf{C}$.

The next lemma is given here to motivate the restriction to torsion free $\mathbf{C}[\chi]$-modules in the subsequent working in this section. This lemma is not used in these notes but is a crucial step in the embedding of the algebra $\mathbf{C}[[\chi]]$ of all formal power series in one indeterminate into V, which is the algebra obtained by adjoining an identity to the convolution algebra $L_1[0, 1]$ (Allan [1, Lemma 1]).

8.4. Let A be a commutative Banach algebra, let x be in A, let $T : A \to A : a \mapsto ax$, and let $J = \cap\{x^n A : n$ is a positive integer$\}$. If $(x^{m+1}A)^- \supseteq x^m A$ for some positive integer m, then T is a linear isomorphism on J.

Proof. We begin by proving that T is one-to-one. Let $y = x^{m+1}a$ be in $x^{m+1}A$ and let $Ty = 0$. Then $x^{m+2}a = 0$ so that $a(x^{m+2}A) = \{0\}$. Because $(x^m A)^- = (x^{m+1}A)^- = (x(x^m A)^-)^- = (x(x^{m+1}A)^-)^- = (x^{m+2}A)^-$, we have $ax^m A = \{0\}$. Thus $y = ax^m x = 0$.

Clearly $TJ \subseteq J$. Let w be in J. Then $w = x^n y_n$ for some sequence (y_n) and all positive integers n. If $k, n > m + 2$, then $x^{n-1}y_n$ and $x^{k-1}y_k$ are in $x^{m+1}A$ and so are equal since $T(x^{n-1}y_n) = x^n y_n = w = x^k y_k = T(x^{k-1}y_k)$. Thus the sequence $(x^{n-1}y_n)$ is constant for $n > m + 2$. This constant z is in J, and $Tz = w$. This completes the proof.

The next lemma is a standard result in field theory where it is usually given in a stronger form involving separable extensions (for example, see [50]).

8.5. Lemma. *Let* $\mathcal{G}$ *be a subfield of a field* $\mathcal{F}$ *of characteristic* 0, *let* X *be an* $\mathcal{F}$ *linear space, and let* Γ *be a transcendence basis for* $\mathcal{F}$ *over* $\mathcal{G}$. *If* δ_0 *is a derivation from* $\mathcal{G}$ *into* X, *then* δ_0 *may be extended to a derivation* δ *from* $\mathcal{F}$ *into* X *and* δ *may be arbitrarily defined on* Γ.

Proof. Let $\mathcal{I}$ be a subfield of $\mathcal{F}$ containing $\mathcal{G}$ and suppose δ_0 has been extended to a derivation δ_1 from $\mathcal{I}$ into X. Let f be in $\mathcal{F}$. If we show how to extend δ_1 to the field generated by $\mathcal{I}$ and f, then an application of Zorn's Lemma will complete the proof. If f is algebraic over $\mathcal{I}$, we let

$$F(\chi) = \chi^m + g_{m-1}\chi^{m-1} + \ldots + \chi g_1 + g_0$$

be the minimal polynomial of f over $\mathcal{I}$. We define $\delta(f)$ by

$$\delta(f) = -\Big(\sum_1^m r g_r f^{r-1}\Big)^{-1} \sum_0^m f^r \delta_1(g_r)\ ;$$

observe that $\sum_1^m r g_r f^{r-1} \neq 0$ because F is the minimal polynomial of f. Now the field generated by $\mathcal{I}$ and f is isomorphic to $\mathcal{I}[\chi]/(F(\chi))$, where $(F(\chi))$ is the principal ideal in $\mathcal{I}[\chi]$ generated by $F(\chi)$. We regard X as an $\mathcal{I}[\chi]$-module by defining $G(\chi).x = G(f)x$ for all $G(\chi)$ in $\mathcal{I}[\chi]$. We now define $D : \mathcal{I}[\chi] \to X$ by

$$D(G(\chi)) = \sum_1^n j h_j f^{j-1} \delta(f) + \sum_0^n f^j \delta(h_j)$$

where $G(\chi) = \sum_0^n h_j \chi^j$. A direct calculation shows that D is a derivation from $\mathcal{I}[\chi]$ into X, and $D((F(\chi))) = \{0\}$ because $F(f) = 0$ and $D(F(\chi)) = 0$ by choice of F and $\delta(f)$. Thus D gives rise to a derivation from $\mathcal{I}[\chi]/((F(\chi)))$ into X, and by the isomorphism this gives a derivation from the field generated $\mathcal{I}$ and f into X.

If f is transcendental over $\mathcal{I}$ we may choose $\delta(f)$ arbitrarily, and extend δ to the field generated by I and f by the derivation property $\delta(gh) = g\delta(h) + h\delta(g)$. This completes the proof.

We convert the above result into a more convenient form for our use in Theorem 8.7. Let X be a divisible torsion free module over an integral domain D, and let $\mathcal{F}$ be the field of fractions of D. Then X may be regarded as an $\mathcal{F}$ linear space by defining $gh^{-1}x = gy$, where $hy = x$ and g and h are in D with $h \neq 0$. The torsion free divisibility is used to choose the unique y.

8.6. Lemma. Let X be a divisible torsion free module over an integral domain D, and let P be a subring of D. If δ_0 is a derivation from P into X, then δ_0 may be extended to a derivation δ from D into X. If Γ is a transcendental subset of D over P, then $\delta : \Gamma \to X$ may be arbitrarily chosen.

Proof. Let $\mathcal{F}$ be the field of fractions of D, and let $\mathcal{G}$ be the subfield generated by P. We regard X as an $\mathcal{F}$ linear space. We extend δ_0 to a derivation δ_1 from $\mathcal{G}$ into X by defining $\delta_1(gh^{-1}) = h^{-1}\delta_0(g) - gh^{-2}\delta_0(h)$ for all g and h in P with $h \neq 0$. Then δ_1 is a derivation from $\mathcal{G}$ into X. Lemma 8.5 enables us to extend δ_1 to a derivation δ from $\mathcal{F}$ into X. The restriction of δ to D is the required derivation.

The difficulty in applying the above result is in constructing divisible modules over the domains D that arise in analysis. The Mittag-Leffler Lemma (Lemma 1.7) has countability built into its hypotheses, and for this reason quasinilpotent operators are used here. H. G. Dales [28, §3] used the group of units (invertible elements) of the domain D to show that a P-divisible module is D-divisible. The P-divisibility is obtained from Lemma 1.7. This is the motivation for the hypotheses of the following theorem. If A is a unital ring, let A^{-1} denote the group of units in A.

8.7. Theorem. Let D be an integral domain, let P be a subring of D such that $D = D^{-1}P$, and let X be a unital D-module. If the maximal P-divisible submodule Z of X is torsion free, then a derivation δ_0 from P into Z may be extended to a derivation δ from D into Z, and δ may be arbitrarily defined on a subset of D that is transcendental over P.

Proof. The theorem will follow from Lemma 8.6 once we prove that Z is D divisible. If f is in D^{-1}, then fZ and $f^{-1}Z$ are P-divisible submodules of X so that $fZ \subseteq Z$ and $f^{-1}Z \subseteq Z$. Hence $fZ = Z$. Therefore $hZ = Z$ for all non-zero h in D because $D = D^{-1}P$. The torsion free property also follows from the relation $D = D^{-1}P$, and the proof is complete.

We apply this theorem to obtain [28, Theorem 2]. There are many continuous linear operators T with the properties required in the example. Any quasinilpotent operator T on a Banach space X such that $(TX)^{-} = X$ and T is one-to-one is a suitable candidate (see §3, [68], [1], or [28]).

8. 8. Example. Let $\mathcal{O}$ be the algebra of germs of analytic functions on the closure Δ^{-} of the open unit disc $\Delta = \{\mu \in \mathbf{C} : |\mu| < 1\}$, and let $\mathcal{P}$ be the subalgebra of $\mathcal{O}$ of polynomials in one variable. Let T be a continuous linear operator on a Banach space X with spectral radius less than 1, and regard X as an $\mathcal{O}$-module by $f.x = f(T)x$, where $f(T)$ is given by the analytic functional calculus (or convergent power series for f about 0). Now $\mathcal{O} = \mathcal{O}^{-1}\mathcal{P}$. Let Γ be a non-empty subset of $\mathcal{O}$ transcendental over $\mathcal{P}$ (see [50]). If the maximal T-divisible subspace X_{∞} of X is non-zero and torsion free, then the hypotheses of Theorem 8. 7 are satisfied. Thus there is a derivation δ from $\mathcal{O}$ into X_{∞} that is zero on $\mathcal{P}$ and arbitrarily defined on Γ. The freedom of choice of δ on Γ enables us to make δ behave badly on Banach algebras embedded in $\mathcal{O}$.

8. 9. Theorem. Let $A(\Delta)$ be the uniform algebra of complex functions continuous in the closed disc Δ^{-} and analytic in the open disc Δ. Then there is a Banach $A(\Delta)$-module X and a discontinuous derivation from $A(\Delta)$ into X.

Proof. Let $\theta : A(\Delta) \to \mathcal{O}$ be defined by $(\theta f)(\mu) = f(\mu/2)$ for all μ in 2Δ and all f in $A(\Delta)$. Then θ is a monomorphism from $A(\Delta)$ into $\mathcal{O}$, and $\theta(A(\Delta))$ contains an element of $\mathcal{O}$ that is transcendental over $\mathcal{P}$ (e.g. $\exp z$). The $\mathcal{O}$-module X of Example 8. 8 may be regarded as an $A(\Delta)$ module by identifying $A(\Delta)$ and $\theta(A(\Delta))$. Thus there is a derivation δ from $A(\Delta)$ into X annihilating the polynomials and non-zero on some element of $A(\Delta)$. Because the polynomials are dense in $A(\Delta)$, the derivation δ is not continuous. To complete the proof we have only to show that X is a Banach $A(\Delta)$-module. This follows from the continuity of the single variable analytic functional calculus or may be obtained directly as follows. Since the spectral radius of T is less than one, there is a constant M such that $\|T^n\| \leq M$ for all positive integers n. If $f(\mu) = \sum_0^\infty \alpha_n \mu^n$ is in $A(\Delta)$, then

$$|\alpha_n| = |f^{(n)}(0)|/n! \leq \|f\|/r^n$$

for all $0 < r < 1$ and all positive integers n by the Cauchy estimates. Hence $|\alpha_n| \leq \|f\|$ for all n. We now obtain

$$\|f.x\| = \|f(T/2)x\| \leq \sum_0^\infty |\alpha_n| . \|T^n\| . 2^{-n} . \|x\| \leq 2M\|f\| . \|x\|.$$

This completes the proof.

8.9. Remarks. (a) Theorem 8.7 may also be applied to certain Frechet algebras. For example, G. R. Allan's embedding [1] of $\mathbf{C}[[\chi]]$ into $V = L^1[0, 1] \oplus \mathbf{C}1$, where $L^1[0, 1]$ has convolution multiplication, turns V into a $\mathbf{C}[[\chi]]$-module. This module has a divisible torsion free $\mathbf{C}[\chi]$-submodule, which was used in the existence of the embedding. By applying Theorem 8.7 there exists a derivation from $\mathbf{C}[[\chi]]$ into V which is zero on $\mathbf{C}[\chi]$ and arbitrary on a non void subset of $\mathbf{C}[[\chi]]$ transcendental over $\mathbf{C}[\chi]$. This derivation may be chosen with sufficient freedom to be discontinuous from $\mathbf{C}[[\chi]]$ (with either its natural Frechet topology or the norm topology induced by V and θ) into V.

(b) There are several papers relating the continuity of derivations and homomorphisms [65], [79], [81], [83].

9. Homomorphisms between $\mathbf{C}(\Omega)$-modules

In this section we give some of the technical results we shall require to study homomorphisms from C*-algebras. We consider modules over $\mathbf{C}(\Omega)$ as this will permit us to use these lemmas for $\mathbf{C}(\Omega)$-homomorphisms (§10) and homomorphisms from non-commutative C*-algebras

(§12). Throughout this section Ω will denote a compact Hausdorff space, X a (left) Banach $C(\Omega)$-module satisfying $\|fx\| \le \|f\| . \|x\|$ for all x in X and f in $C(\Omega)$, and Y a $C(\Omega)$-module with continuous module operations. We are assuming that there is a homomorphism from $C(\Omega)$ into the Banach algebra of continuous linear operators on Y. We let T be a $C(\Omega)$-module homomorphism from X into Y.

The following lemma is similar to [7, Theorem 2.1] but is weaker than this theorem. It is, however, strong enough to obtain the main results of Bade and Curtis [7] (Theorem 10.4).

9.1. Lemma. *Let* (f_n) *be a sequence of non-zero elements in* $C(\Omega)$ *such that* $f_m f_n = 0$ *for all* $m \neq n$. *Then* $f_n^2 T$ *is continuous for all but a finite number of* n, *and the set* $\{\|f_n^3 T\| / \|f_n\|^3 : f_n^2 T$ *is continuous*$\}$ *is bounded.*

Proof. Suppose that $f_n^2 T$ is discontinuous for an infinite number of n, then, by passing to a subsequence, we may assume that $f_n^2 T$ is discontinuous for all n. Choose a sequence (x_n) from X so that $\|x_n\| \le 2^{-n}\|f_n\|^{-1}$ and $\|f_n^2 T(x_n)\| \ge nK_n$, where K_n is the norm of the linear operator from Y into Y defined by $y \mapsto f_n y$. Let $z = \sum_1^\infty f_n x_n$. Then

$$K_n\|Tz\| \ge \|f_n Tz\| = \|T(f_n z)\| = \|T(f_n^2 x_n)\| \ge nK_n.$$

Because $f_n^2 T$ is discontinuous, K_n is non-zero and $\|Tz\| \ge n$ for all positive integers n. This proves that $f_n^2 T$ is continuous for all but a finite number of n.

We now suppose that $f_n^2 T$ is continuous for all n, and that $\{\|f_n^3 T\| / \|f_n\|^3\}$ is unbounded. We may assume that $\|f_n\| = 1$ for all n. Form (f_n) into infinitely many disjoint subsequences (f_{ij}) so that $\|f_{ij}^3 T\| > 4^j . j$ for all i and j. Choose x_{ij} in X with $\|x_{ij}\| = 1$ and $\|f_{ij}^3 T(x_{ij})\| > 4^j . j$ for all i and j. Let $h_n = \sum_j f_{nj} 2^{-j}$ for all n. Then $h_n h_m = 0$ for all $m \neq n$, and

$$\|h_n^2 T(f_{nj} x_{nj})\| = \|f_{nj}^3 T(x_{nj})\| 4^{-j} > j \ge j\|f_{nj} x_{nj}\|$$

for all n and j. Hence $h_n^2 T$ is discontinuous for all n. This contradicts the first half of the lemma.

9.2. Definition. If F is a subset of Ω, we define J(F) and Ker(F) by

$$J(F) = \{f \in \mathbf{C}(\Omega) : f \text{ is zero in a neighbourhood of } F\},$$

and

$$\mathrm{Ker}(F) = \{f \in \mathbf{C}(\Omega) : f(F) = \{0\}\}.$$

Then Ker(F) and J(F) are ideals in $\mathbf{C}(\Omega)$.

If τ is an ideal in $\mathbf{C}(\Omega)$ with $\{\lambda \in \Omega : \tau(\lambda) = \{0\}\} = F$, then $J(F) \subseteq \tau \subseteq \mathrm{Ker}(F)$. The set $\{\lambda \in \Omega : \tau(\lambda) = \{0\}$ is called the _hull_ of τ.

9.3. Theorem. _Let_ $\tau = \{f \in \mathbf{C}(\Omega) : f\mathfrak{S}(T) = \{0\}\}$. _Then_

(i) τ _is an ideal in_ $\mathbf{C}(\Omega)$ _with finite hull_ F, _and_

(ii) T _is continuous on the linear subspace_ J(F). X _of_ X.

Proof. (i) Since Y is a $\mathbf{C}(\Omega)$-module, τ is an ideal in $\mathbf{C}(\Omega)$. Suppose that the hull F of τ is infinite. Then we can construct a sequence $\{U_n\}$ of open subsets of Ω such that $U_j \cap U_k = \emptyset$ for all $j \neq k$ and $U_k \cap F \neq \emptyset$ for all k. Let λ_k be in $U_k \cap F$. We choose f_k in $\mathbf{C}(\Omega)$ so that $f_k(\lambda_k) \neq 0$ and $f_k(\Omega \backslash U_k) = \{0\}$ for all k. Then $f_k^2 T$ is discontinuous for all k because f_k^2 is not in τ. This contradicts Lemma 8.1 because $f_k f_j = 0$ if $j \neq k$.

(ii) Note that $J(F).X = \{fx : f \in J(F),\ x \in X\}$ is a linear subspace of X because if f_1, f_2 are in J(F) then there is an f in J(F) so that $ff_1 = f_1$ and $ff_2 = f_2$ which implies that $f_1x_1+f_2x_2=f(f_1x_1+f_2x_2)$. Suppose that the restriction of T to J(F). X is discontinuous. If f and g in J(F) satisfy $\|f\| = 1$ and $fg = g$, then

$$\|f^3T\| \geq \|f^3T(gx)\|/\|gx\| = \|T(gx)\|/\|gx\|.$$

Let (g_nx_n) be a sequence of elements of unit norm from J(F). X such that the sequence $(\|T(g_nx_n)\|)$ is unbounded. By induction we choose a sequence (f_n) from J(F) so that $\|f_n\| = 1$, $f_nf_{n-1} = f_{n-1}$, and

$f_n g_n = g_n$. We may also choose f_n so that it is one at some point at which f_{n-1} is zero. Hence the sequence $(\|f_n^3 T\|)$ is unbounded. We choose a subsequence (k_n) from (f_n) so that $\|k_n^3 T\| \geq n + \|k_{n-1}^3 T\|$ for all n. Then $\|(k_n^3 - k_{n-1}^3)T\| \geq n$. Because $k_n k_{n-1} = k_{n-1}$, we have $k_n^3 - k_{n-1}^3 = (k_n - k_{n-1})^3$. Also $\|k_n - k_{n-1}\| \geq 1$ because there is a point where k_n is 1 and k_{n-1} is 0. Let $h_n = k_{2n} - k_{2n-1}$ for all n. Then the set $\{\|h_n^3 T\| / \|h_n\|^3\}$ is unbounded. If $n > m$, then

$$\begin{aligned} h_n h_m &= (k_{2n} - k_{2n-1})k_{2m} - (k_{2n} - k_{2n-1})k_{2m-1} \\ &= (k_{2m} - k_{2m}) - (k_{2m-1} - k_{2m-1}) = 0. \end{aligned}$$

These properties of the sequence (h_n) contradict Lemma 9.1.

9.4. **Corollary.** If Y is a Banach $C(\Omega)$-module, then T is continuous on the Banach $C(\Omega)$-submodule Ker(F).X, where F is the hull of τ.

Proof. Let M be the norm of the restriction of T to J(F).X, and let $\|f.y\| \leq N\|f\|.\|y\|$ for all f in $C(\Omega)$ and y in Y. If f is in Ker(F) and x is in X, then there is a g in J(F) such that $\|f - g\| < \varepsilon$, where ε is chosen later. Hence

$$\begin{aligned} \|T(fx)\| &\leq N.\|f - g\|.\|Tx\| + \|T(gx)\| \\ &\leq \varepsilon.N.\|Tx\| + M.\|gx\| \\ &\leq \varepsilon.N.\|Tx\| + M.\varepsilon.\|x\| + M.\|fx\| \\ &\leq (M + 1)\|fx\| \end{aligned}$$

provided $\varepsilon.N.\|Tx\| + M.\varepsilon.\|x\| \leq \|fx\|$. Cohen's Factorization Theorem [18], [49] implies that Ker(F).X is a Banach subspace of X, and it is a submodule because Ker(F) is an ideal in $C(\Omega)$. This completes the proof.

9.5. **Remarks.** Various versions of Theorem 9.3 occur in [24], [78], [117]. Lemma 9.1 is essentially a special case of Bade and Curtis [7, Theorem 2.1]. The results of Sections 9, 10, and 12 do not depend on Lemma 1.6 or Theorem 2.3. Something similar to Lemma 9.1 may be deduced from Lemma 1.6 but the only proof I know is harder than that of 9.1. This suggests that 1.6 (or 2.3) can be strengthened, or that these are the wrong technical results.

10. Homomorphisms from $C(\Omega)$

We begin this section with Kaplansky's Theorem on monomorphisms from $C(\Omega)$ into a Banach algebra [72], [103]. A corollary (10.2) of this will be required as a lemma in the proof of Theorem 10.3, which is the main result of Bade and Curtis's paper [7].

10.1. Theorem. Let Ω be a compact Hausdorff space. If θ is a monomorphism from $C(\Omega)$ into a Banach algebra B, then $\|\theta(f)\| \geq \|f\|$ for all f in $C(\Omega)$.

Proof. We may assume that θ is onto a dense subalgebra of B. Thus B is commutative and has an identity. Let Φ denote the carrier space of B, and let $\theta^* : \Phi \to \Omega$ be defined by $f(\theta^*(\psi)) = \psi\theta(f)$ for all ψ in Φ and f in $C(\Omega)$ - we are using the result that the carrier space of $C(\Omega)$ is homeomorphic to Ω under the natural embedding of Ω into the carrier space of $C(\Omega)$. The map θ^* from Φ into Ω is continuous, and hence $\theta^*\Phi$ is compact, so closed, in Ω. Suppose θ^* is not onto Ω, and let λ_0 be in $\Omega\backslash\theta^*\Phi$. We choose h in $C(\Omega)$ such that h is 1 in a neighbourhood of λ_0 and $h(\theta^*\Phi) = \{0\}$, and a non-zero g in $C(\Omega)$ such that $hg = g$. Since $h(\theta^*\Phi) = \{0\}$ it follows that $(\theta h)\hat{}(\Phi) = \{0\}$ so that θh is in the radical of B. From $(1 - \theta(h))\theta(g) = 0$ we obtain $\theta(g) = 0$. This contradicts θ being a monomorphism. Hence $\theta^*\Phi = \Omega$. If f is in $C(\Omega)$, then $\|f\| = \sup|f(\Omega)| = \sup|f(\theta^*\Phi)| = \sup|(\theta f)(\Phi)| =$ spectral radius of $\theta f \leq \|\theta f\|$. This completes the proof.

10.2. Corollary. Let Ω be a compact Hausdorff space. If μ is a continuous homomorphism from $C(\Omega)$ into a Banach algebra B, then $\mu(C(\Omega))$ is closed in B.

Proof. Let $\theta : C(\Omega)/\mathrm{Ker}\,\mu \to B$ be defined by $\theta(f+\mathrm{Ker}\,\mu) = \mu(f)$. Then θ is a continuous monomorphism from $C(\Omega)/\mathrm{Ker}\,\mu$ into B, and $C(\Omega)/\mathrm{Ker}\,\mu$ is a commutative C^*-algebra. Thus $C(\Omega)/\mathrm{Ker}\,\mu$ is isometrically isomorphic to the Banach algebra of continuous functions on its carrier space. By Theorem 10.1 θ is norm increasing, and so the image of θ, which is $\mu(\mathbf{C}(\Omega))$, is complete and closed in B.

10.3. Theorem. Let Ω be a compact Hausdorff space, and let θ be a homomorphism from $\mathbf{C}(\Omega)$ onto a dense subalgebra of a Banach algebra B with radical R. Let $\tau = \{f \in \mathbf{C}(\Omega) : \theta(f)\mathfrak{S}(\theta) = \{0\}\}$. Then

(i) τ is an ideal in $\mathbf{C}(\Omega)$ with finite hull $F = \{\lambda_1, \ldots, \lambda_n\}$, say;

(ii) there is a continuous homomorphism μ from $\mathbf{C}(\Omega)$ into B that coincides with θ on a dense subalgebra of $\mathbf{C}(\Omega)$ containing $\tau^-.\tau$;

(iii) the range of μ is closed in B and $B = \mu(\mathbf{C}(\Omega)) \oplus R$ as Banach spaces;

(iv) $R = \mathfrak{S}(\theta)$ and $\mu(\tau^-).R = \{0\}$;

(v) $\operatorname{Ker}\mu = (\operatorname{Ker}\theta)^- = \theta^{-1}\mathfrak{S}(\theta)$;

(vi) if $\nu = \theta - \mu$, then ν is a homomorphism from Ker(F) onto a dense subalgebra of $\mathfrak{S}(\theta)$, ν annihilates $\tau^-.\tau$, and $\mathfrak{S}(\nu) = \mathfrak{S}(\theta)$; and

(vii) there are linear operators $\nu_1, \ldots, \nu_n$ from $\mathbf{C}(\Omega)$ into B such that

(a) $\nu = \nu_1 + \ldots + \nu_n$,

(b) $R = R_1 \oplus \ldots \oplus R_n$ where $R_j = (\nu_j(\mathbf{C}(\Omega)))^-$,

(c) $R_jR_k = \{0\}$ if $j \neq k$,

(d) the restriction of ν_j to $\operatorname{Ker}(\{\lambda_j\})$ is a homomorphism into B, and

(e) $\nu_j(J(\{\lambda_j\})) = \{0\}$.

Proof. (i) This follows from Theorem 9.3(i) by regarding $\mathbf{C}(\Omega)$ as a Banach $\mathbf{C}(\Omega)$-module and B as a $\mathbf{C}(\Omega)$-module by $f.b = \theta(f).b$ for all f in $\mathbf{C}(\Omega)$ and b in B.

(ii) Choose $f_1, \ldots, f_n$ in $\mathbf{C}(\Omega)$ so that $f_jf_k = 0$ if $j \neq k$ and f_j is equal to 1 in a neighbourhood of λ_j for each j. Let $D = \mathbf{C}f_1 \oplus \ldots \oplus \mathbf{C}f_n \oplus J(F)$. Then D is a subalgebra of $\mathbf{C}(\Omega)$ because $f_j^2 = f_j$ modulo J(F). Let μ_0 be the restriction of θ to D. Then μ_0 is a continuous homomorphism from D into B because θ is continuous on J(F) (Theorem 9.3(ii)) and the sum $\mathbf{C}f_1 \oplus \ldots \oplus \mathbf{C}f_n \oplus J(F)$ is a normed direct sum. Since D is dense in $\mathbf{C}(\Omega)$ and B is complete we may lift μ_0 by continuity to a continuous homomorphism μ from $\mathbf{C}(\Omega)$ into B. Let f be in τ^-, let g be in τ, and let b_n in J(F) converge to f, which is in $\mathrm{Ker}(F) = J(F)^-$. Then

$$\theta(fg) = \lim \theta(gb_n) = \lim \mu(gb_n) = \mu(gf).$$

(iii) By Corollary 10.2 $\mu(\mathbf{C}(\Omega))$ is closed in B, and is algebraically isomorphic to $\mathbf{C}(\Omega)/\mathrm{Ker}\ \mu$. Hence $\mu(\mathbf{C}(\Omega))$ is semisimple, and therefore

$$\mu(\mathbf{C}(\Omega)) \cap \mathbf{R} = \{0\}. \tag{1}$$

Let $Q : B \to B/\mathfrak{S}(\theta)$ be the natural quotient map. Then $Q\theta$ and $Q\mu$ are continuous and agree on a dense subalgebra of $\mathbf{C}(\Omega)$ so $Q\theta = Q\mu$. Further $Q\theta = Q\mu$ is a continuous homomorphism from $\mathbf{C}(\Omega)$ onto a dense subalgebra of $B/\mathfrak{S}(\theta)$. By Theorem 10.2 $Q(\mu(\mathbf{C}(\Omega)))$ is equal to $B/\mathfrak{S}(\theta)$. Therefore

$$B = \mu(\mathbf{C}(\Omega)) + \mathfrak{S}(\theta). \tag{2}$$

If ω is a character on B, then $\omega\theta$ is a character on $\mathbf{C}(\Omega)$ so is continuous. Thus $\mathfrak{S}(\theta)$ is contained in $\mathrm{Ker}\ \omega$, and hence

$$\mathfrak{S}(\theta) \subseteq \mathbf{R}. \tag{3}$$

(1), (2), and (3) imply that

$$B = \mu(\mathbf{C}(\Omega)) \oplus \mathfrak{S}(\theta) \quad \text{and} \quad \mathfrak{S}(\theta) = \mathbf{R}.$$

(iv) If f is in $\tau^-.\tau$, then $\theta(f)\mathfrak{S}(\theta) = \mu(f)\mathfrak{S}(\theta) = \{0\}$ by (ii) and the definition of τ. The continuity of μ and $\tau^- = (\tau^-.\tau)^-$ gives $\mu(\tau^-)\mathfrak{S}(\theta) = \{0\}$.

(vi) By the proof of (iii) $Q\nu = 0$ so that $\nu : \mathbf{C}(\Omega) \to \mathfrak{S}(\theta)$. Let f and g be in Ker(F). Then

$$\begin{aligned} \mu(fg) + \nu(fg) = \theta(fg) &= \theta(f)\theta(g) \\ &= (\mu(f) + \nu(f))(\mu(g) + \nu(g)) \\ &= \mu(fg) + \nu(f)\nu(g) \end{aligned}$$

since $\mu(\text{Ker}(F)). \mathfrak{S}(\theta) = \{0\}$. Thus ν is a homomorphism from $\text{Ker}(F)$ into $\mathfrak{S}(\theta)$. Because $\theta = \mu + \nu$ and μ is continuous, $\mathfrak{S}(\theta) = \mathfrak{S}(\nu)$. Hence ν is onto a dense subalgebra of $\mathfrak{S}(\theta)$.

(v) Clearly $\text{Ker}\ \theta \subseteq \text{Ker}\ \mu$, and so $(\text{Ker}\ \theta)^- \subseteq \text{Ker}\ \mu$. If h is in $\text{Ker}\ \mu$, then there are f and g in $\text{Ker}\ \mu$ such that $h = f.g$ because $\text{Ker}\ \mu$ is a Banach algebra with a bounded approximate identity and Cohen's Factorization Theorem [25] applies (or use the method of Theorem 11. 7). By (ii) there is a sequence (k_n) in $\tau^-.\tau$ such that $\nu(k_n) = 0$ for all n, and $k_n \to f$ as $n \to \infty$. Then $k_n g$ converges to h, and

$$\theta(k_n g) = \theta(k_n)\theta(g) = (\mu(k_n) + \nu(k_n)).\ \nu(g) = \nu(k_n)\nu(g) = 0.$$

Hence $\text{Ker}\ \mu \subseteq (\text{Ker}\ \theta)^-$.

Let h be in $\mathbf{C}(\Omega)$. Then $\theta(h) = \mu(h) + \nu(h)$ is in $\mathfrak{S}(\theta)$ if and only if $\mu(h) = 0$. Therefore $\theta^{-1}\ \mathfrak{S}(\theta) = \text{Ker}\ \mu$.

(vii) Using the f_j defined in the proof of (ii) we let $\nu_j(g) = \nu(f_j g)$ for $j = 1, \ldots, n$. By (iv), we have $\nu((1 - f_1 \ldots - f_n)k) = 0$ for all k in $\mathbf{C}(\Omega)$. Hence $\nu(k) = \nu_1(k) + \ldots + \nu_n(k)$. Thus (a) is proved. If f and g are in $\text{Ker}(\{\lambda_j\})$, then ff_j and gf_j are in $\text{Ker}(F)$, and $(f_j^2 - f_j)fg$ is in $J(F)$ so that

$$\nu_j(f)\nu_j(g) - \nu_j(fg) = \nu((f_j^2 - f_j)fg) = 0.$$

Hence the restriction of ν_j to $\text{Ker}(\{\lambda_j\})$ is a homomorphism. Further $\nu_j(J(\{\lambda_j\})) = \{0\}$ because

$$\nu_j(J(\{\lambda_j\})) = \nu(f_j J(\{\lambda_j\})) \subseteq \nu(J(F)) = \{0\}.$$

This proves (d) and (e).

To prove (b) and (c) we shall use the equality $\mu(f_j)\nu_j(g) = \nu_j(g)$ for all g in $\mathbf{C}(\Omega)$ and $j = 1, \ldots, n$. Since $\mu(f_j) = \theta(f_j)$, we have

$$\begin{aligned} \mu(f_j)\nu_j(g) &= \mu(f_j)\{\theta(f_j g) - \mu(f_j g)\} \\ &= \theta(f_j^2 g) - \mu(f_j^2 g) = \nu(f_j^2 g) = \nu(f_j g) = \nu_j(g). \end{aligned}$$

Hence $\mu(f_j)\nu_k(g) = \mu(f_j f_k)\nu_k(g) = 0$ if $j \neq k$. Therefore $\{\mu(f_j) : j=1, \ldots, n\}$

is a set of orthogonal idempotent operators on $\nu(\mathbf{C}(\Omega))$, and thus on $\nu(\mathbf{C}(\Omega))^{-} = R$. This shows that $R = \mu(f_1)R \oplus \ldots \oplus \mu(f_n)R$. Now (b) is completed by observing that $\mu(f_j)R = \nu_j(\mathbf{C}(\Omega))^{-}$. Equation (c) follows from $R_j R_k = \mu(f_j f_k)R.R = \{0\}$ if $j \neq k$. The proof is complete.

In Remark 11.8 we show that 10.3(ii) may be strengthened by replacing $\tau^{-}.\tau$ by τ in (ii).

10.4. Corollary. *Let* Ω *be a compact Hausdorff space. There is a discontinuous homomorphism from* $\mathbf{C}(\Omega)$ *into a Banach algebra if and only if there is a* λ *in* Ω *and a non-zero homomorphism from* $\mathrm{Ker}(\{\lambda\})$ *into a radical Banach algebra that annihilates* $J(\{\lambda\})$.

11. Homomorphisms into radical Banach algebras

In this section we investigate homomorphisms from $\mathbf{C}_0(\Psi)$ into radical Banach algebras, where Ψ is a locally compact Hausdorff space. The motivation for this work is Corollary 10.4, which shows that there is a discontinuous homomorphism from some $\mathbf{C}(\Omega)$ into a Banach algebra if and only if there is a discontinuous homomorphism from some $\mathbf{C}_0(\Psi)$ into a radical Banach algebra. The main result of this section is Theorem 11.7, and the crucial technical results are Lemmas 11.1 and 11.3 both of which depend on Lemma 1.6.

Let $\mathbf{P}$ denote the *additive* semigroup of positive rational numbers (the proofs work if it is the set of positive elements of any dense additive subgroup of $\mathbf{R}$).

11.1. Lemma. *Let* X *and* Y *be Banach spaces, and let* $\alpha \mapsto T(\alpha)$ *and* $\alpha \mapsto R(\alpha)$ *be homomorphisms from* $\mathbf{P}$ *into the multiplicative semigroups of* $\mathcal{L}(X)$ *and* $\mathcal{L}(Y)$, *respectively. If* S *is a linear operator from* X *into* Y *satisfying* $ST(\alpha) = R(\alpha)S$ *for all* α *in* $\mathbf{P}$, *then* $(R(\alpha)\mathfrak{S})^{-} = (R(\beta)\mathfrak{S})^{-}$ *for all* α *in* β *in* $\mathbf{P}$.

Proof. If α and β are in $\mathbf{P}$, and if $\alpha < \beta$, then $(R(\beta)\mathfrak{S})^{-} = (R(\alpha)R(\beta - \alpha)\mathfrak{S})^{-} \subseteq (R(\alpha)\mathfrak{S})^{-}$ because $(R(\gamma)\mathfrak{S})^{-} \subseteq \mathfrak{S}$ for all γ in $\mathbf{P}$. We now show that if α and β in $\mathbf{P}$ with $\alpha < \beta$ satisfy $(R(\alpha)\mathfrak{S})^{-} = (R(\beta)\mathfrak{S})^{-}$, then $R(\gamma)\mathfrak{S})^{-} = (R(\alpha)\mathfrak{S})^{-}$ for all $\gamma > \alpha$. We begin by proving that

$$(R(\alpha)\mathfrak{S})^- = (R(\alpha + n(\beta - \alpha))\mathfrak{S})^-$$

for all positive integers n. This is true for $n = 1$ by supposition so assume that it is true for n. Then

$$\begin{aligned}\{R(\alpha + (n+1)(\beta - \alpha))\mathfrak{S}\}^- &= \{R(\beta - \alpha)[R(\alpha + n(\beta - \alpha))\mathfrak{S}]^-\}^- \\ &= \{R(\beta - \alpha)[R(\alpha)\mathfrak{S}]^-\}^- \\ &= \{R(\beta)\mathfrak{S}\}^- = \{R(\alpha)\mathfrak{S}\}^-.\end{aligned}$$

If γ is in **P** with $\alpha < \gamma$, then there is a positive integer n so that $\alpha < \gamma < \alpha + n(\beta - \alpha)$. Thus

$$(R(\alpha)\mathfrak{S})^- \supseteq (R(\gamma)\mathfrak{S})^- \supseteq \{R(\alpha + n(\beta - \alpha))\mathfrak{S}\}^- = (R(\alpha)\mathfrak{S})^-.$$

This proves the constancy of $(R(\gamma)\mathfrak{S})^-$ beyond α provided that $(R(\alpha)\mathfrak{S})^- = (R(\beta)\mathfrak{S})^-$ with $\alpha < \beta$.

Now suppose that α and β in **P** satisfy $\alpha < \beta$ and $(R(\alpha)\mathfrak{S})^- \neq (R(\beta)\mathfrak{S})^-$. Choose a sequence (α_n) from **P** so that $\alpha_1 + \ldots + \alpha_n < \alpha$ for all n, and let $T_j = T(\alpha_j)$ and $R_j = R(\alpha_j)$ for all positive integers j. Then $(R_1 \ldots R_n \mathfrak{S})^- \supset (R_1 \ldots R_{n+1}\mathfrak{S})^-$ for all n because $\alpha_1 + \ldots + \alpha_n < \alpha_1 + \ldots + \alpha_{n+1} < \alpha < \beta$ and and $(R(\alpha)\mathfrak{S})^- \neq (R(\beta)\mathfrak{S})^-$. This contradicts Lemma 1.6 and completes the proof.

We shall apply the following Lemma when $\mathcal{G}$ is the multiplicative semigroup of real valued non-negative functions on a locally compact Hausdorff space. An additive commutative semigroup $\mathcal{G}$ is said to be <u>divisible</u> if for each g in $\mathcal{G}$ there is a homomorphism $q \mapsto q.g$ from the additive semigroup of positive rationals **P** into $\mathcal{G}$ such that $1.g = g$.

11.2. Lemma. <u>Let</u> X <u>and</u> Y <u>be Banach spaces, let</u> $\mathcal{G}$ <u>be a divisible semigroup, and let</u> $\alpha \mapsto T(\alpha)$ <u>and</u> $\alpha \mapsto R(\alpha)$ <u>be homomorphisms from</u> $\mathcal{G}$ <u>into the multiplicative semigroups of</u> $\mathfrak{L}(X)$ <u>and</u> $\mathfrak{L}(Y)$, <u>respectively. If</u> S <u>is a linear operator from</u> X <u>into</u> Y <u>satisfying</u> $ST(\alpha) = R(\alpha)S$ <u>for all</u> α <u>in</u> $\mathcal{G}$, <u>then</u> $\{(R(\alpha)\mathfrak{S})^- : \alpha \in \mathcal{G}\}$ <u>satisfies the descending chain</u>

condition and $(R(\alpha)\mathfrak{S})^- = (R(n\alpha)\mathfrak{S})^-$ *for all* α *in* $\mathcal{G}$ *and all positive integers* n.

Proof. If α is in $\mathcal{G}$, then there is a homomorphism $q \mapsto q\alpha$ from $\mathbf{P}$ into $\mathcal{G}$ such that $1.\alpha = \alpha$. Using the product of this homomorphism and the homomorphisms $\alpha \mapsto T(\alpha)$ and $\alpha \mapsto R(\alpha)$ we may apply Lemma 11.1. Thus $(R(\alpha)\mathfrak{S})^- = (R(n\alpha)\mathfrak{S})^-$ for all n and α.

If α and β are in $\mathcal{G}$, and if $(R(\alpha)\mathfrak{S})^- \subseteq (R(\beta)\mathfrak{S})^-$, then

$$\begin{aligned}(R(\alpha)\mathfrak{S})^- &= (R(2\alpha)\mathfrak{S})^- = (R(\alpha)(R(\alpha)\mathfrak{S})^-)^- \\ &\subseteq (R(\alpha)(R(\beta)\mathfrak{S})^-)^- = (R(\alpha)R(\beta)\mathfrak{S})^- \subseteq (R(\alpha)\mathfrak{S})^-\end{aligned}$$

so that $(R(\alpha)\mathfrak{S})^- = (R(\alpha)R(\beta)\mathfrak{S})^-$. Suppose that $\alpha_1, \alpha_2, \ldots$ in $\mathcal{G}$ satisfy $(R(\alpha_n)\mathfrak{S})^- \supset (R(\alpha_{n+1})\mathfrak{S})^-$ for all positive integers n. Repeated application of the above observation shows that

$$\begin{aligned}(R(\alpha_n)\mathfrak{S})^- &= (R(\alpha_1)\ldots R(\alpha_n)\mathfrak{S})^- \supset (R(\alpha_1)\ldots R(\alpha_{n+1})\mathfrak{S})^- \\ &= (R(\alpha_{n+1})\mathfrak{S})^-\end{aligned}$$

for all positive integers n. This contradicts Lemma 1.6 when $T_n = T(\alpha_n)$ and $R_n = R(\alpha_n)$, and the proof is complete.

Our second pair of technical results (11.3 and 11.4) are concerned with the construction of a prime ideal from a discontinuous homomorphism. The first part of Lemma 11.3 is related to [7, Theorem 2.1] but the hypothesis and conclusion are weaker than those of that result.

11.3. **Lemma.** *Let* X *and* Y *be Banach spaces, let* $\mathcal{G}$ *be an additive commutative semigroup, let* $\alpha \mapsto T(\alpha)$ *and* $\alpha \mapsto R(\alpha)$ *be homomorphisms from* $\mathcal{G}$ *into the multiplicative semigroups of* $\mathcal{L}(X)$ *and* $\mathcal{L}(Y)$, *respectively, and let* S *be a linear operator from* X *into* Y *satisfying* $ST(\alpha) = R(\alpha)S$ *for all* α *in* $\mathcal{G}$.

(i) *If* (α_n) *and* (β_n) *are sequences in* $\mathcal{G}$ *such that* $R(\beta_n + \alpha_1 + \ldots + \alpha_n)\mathfrak{S} = \{0\}$ *for all* n, *then there is an* N *such that* $R(\beta_n + \alpha_1 + \ldots + \alpha_{n-1})\mathfrak{S} = \{0\}$ *for all* $n \geq N$.

(ii) *If* $R(\delta)S$ *is discontinuous for some* δ *in* $\mathcal{G}$, *then there is a* γ *in* $\mathcal{G}$ *such that* $M = \operatorname{Ker} R(\gamma)$ *does not contain* $\mathfrak{S}$ *and has the property that* α, β *in* $\mathcal{G}$ *and* $R(\alpha+\beta)\mathfrak{S} \subseteq M$ *imply that* $R(\alpha)\mathfrak{S} \subseteq M$ *or*

$R(\beta)\mathfrak{S} \subseteq M$.

Proof. (i) We apply Lemma 1.6 with $T_j = T(\alpha_j)$ and $R_j = R(\alpha_j)$. Then there is an N such that

$$(R(\alpha_1 + \ldots + \alpha_n)\mathfrak{S})^- = (R(\alpha_1 + \ldots + \alpha_{n-1})\mathfrak{S})^-$$

for all $n \geq N$. Thus

$$\begin{aligned} &R(\beta_n + \alpha_1 + \ldots + \alpha_{n-1})\mathfrak{S} \\ &\quad \subseteq R(\beta_n)(R(\alpha_1 + \ldots + \alpha_n)\mathfrak{S})^- \\ &\quad \subseteq (R(\beta_n + \alpha_1 + \ldots + \alpha_n)\mathfrak{S})^- = \{0\} \end{aligned}$$

for all $n \geq N$.

(ii) Suppose there is no such γ in $\mathcal{G}$. We shall construct two sequences (α_n) and (β_n) in $\mathcal{G}$ so that $R(\alpha_1 + \ldots + \alpha_n)\mathfrak{S} \neq \{0\}$, $R(\alpha_1 + \ldots + \alpha_{n-1} + \beta_n)\mathfrak{S} \neq \{0\}$, and $R(\alpha_1 + \ldots + \alpha_n + \beta_n)\mathfrak{S} = \{0\}$ for all n. With $M = \{0\}$ choose α_1 and β_1 in $\mathcal{G}$ so that $R(\alpha_1 + \beta_1)\mathfrak{S} = \{0\}$, $R(\alpha_1)\mathfrak{S} \neq \{0\}$, and $R(\beta_1)\mathfrak{S} \neq \{0\}$. Assume that $\alpha_1, \ldots, \alpha_n, \beta_1, \ldots, \beta_n$ have been chosen. Let $M = \{y \in Y : R(\alpha_1 + \ldots + \alpha_n)y = 0\}$. Then M is a closed linear subspace of Y that is invariant under $R(\alpha)$ for all α in $\mathcal{G}$, because $\mathcal{G}$ is commutative. Also $\mathfrak{S}$ is not contained in M. By supposition there are α_{n+1} and β_{n+1} in $\mathcal{G}$ such that $R(\alpha_{n+1} + \beta_{n+1})\mathfrak{S} \subseteq M$, $R(\alpha_{n+1})\mathfrak{S} \not\subseteq M$, and $R(\beta_{n+1})\mathfrak{S} \not\subseteq M$. Thus $R(\alpha_1 + \ldots + \alpha_{n+1} + \beta_{n+1})\mathfrak{S} = \{0\}$, $R(\alpha_1 + \ldots + \alpha_{n+1})\mathfrak{S} \neq \{0\}$, and $R(\alpha_1 + \ldots + \alpha_n + \beta_{n+1})\mathfrak{S} \neq \{0\}$. This completes the inductive choice of the sequences (α_n) and (β_n). These sequences contradict (i), and the proof is complete.

11.4. Theorem. _Let θ be a discontinuous homomorphism from a commutative Banach algebra A onto a dense subalgebra of a Banach algebra B, and let f be in A. If $\theta(f)$ is a non-nilpotent element in $\mathfrak{S}(\theta)$, then there is a k in A such that $\psi : A \to B/J : a \to \theta(a) + J$ is a discontinuous homomorphism whose kernel is a prime ideal such that $\psi(f)$ is non-zero, where $J = \{b \in B : b\theta(k)\mathfrak{S}(\theta) = \{0\}\}$._

Proof. We shall apply Lemma 11.3(ii) with $X = A$, $Y = B$, $\mathcal{G}$

the multiplicative semigroup of A, T(g) multiplication by g in A, and R(g) multiplication by $\theta(g)$ in B. We now choose S. By Corollary 1.7 applied to the semigroup $\{f^n : n \text{ is a positive integer}\}$ there is a positive integer N such that $(\theta(f)^n \mathfrak{S})^- = (\theta(f)^N \mathfrak{S})^-$ for all $n \geq N$. We let $S(g) = \theta(f)^N \theta(g)$ for all g in A. Then S is discontinuous because $\mathfrak{S}(S) = (\theta(f)^N \mathfrak{S}(\theta))^-$ is non-zero (Lemma 1.3(ii)). Let h be the element in $\mathcal{G}$ given by Lemma 11.3(ii), let $M = \{b \in B : \theta(h)b = 0\}$, let $k = hf^N$, and let $J = \{b \in B : b\theta(k)\mathfrak{S}(\theta) = \{0\}\}$. Then M and J are closed ideals in B and $\mathfrak{S}(S)$ is not contained in M. Because $\mathfrak{S}(S) = (\theta(f)^N \mathfrak{S}(\theta))^-$ it follows that $\theta(h)\theta(f)^N \mathfrak{S}(\theta)$ is non-zero. As $(\theta(f)^N \mathfrak{S}(\theta))^- = (\theta(f)^{N+1} \mathfrak{S}(\theta))^-$ we have $\theta(f)\theta(k)\mathfrak{S}(\theta) \neq \{0\}$ so that $\theta(f)$ is not in J. The choice of h and the definition of J ensure that $\theta(ab)$ in J implies that $\theta(a)$ is in J or $\theta(b)$ is in J. Therefore the kernel of ψ is a prime ideal in A. Now $\mathfrak{S}(\psi) = \mathfrak{S}(Q\theta) = (Q\mathfrak{S}(\theta))^-$, where $Q : B \to B/J$ is the natural quotient map. Therefore $\mathfrak{S}(\psi) = (\mathfrak{S}(\theta)+J)^-/J$ so that ψ will be discontinuous if and only if $\mathfrak{S}(\theta)$ is not contained in J. Because $\theta(f)$ is in $\mathfrak{S}(\theta)\backslash J$ it follows that ψ is discontinuous. This completes the proof.

11.5. Remark. (a) The crucial point about the above theorem is that the prime ideal given is the kernel of a discontinuous homomorphism from A into a Banach algebra. Results from elementary commutative algebra give a prime ideal P in A such that $P \supseteq \operatorname{Ker} \theta$ and f is not in P (see Atiyah and Macdonald [3, Proposition 1.14]).

(b) Note that $\mathfrak{S}(\theta)$ is contained in the radical of B so that $\theta(f)$ is a quasinilpotent non-nilpotent element in B.

(c) The above theorem may be worded in terms of the ideal $\tau = \{g \in A : \theta(g)\mathfrak{S}(\theta) = \{0\}\}$ that occurs in Sections 10 and 12. The conclusion of 11.4 is that there is a k in A such that kf is not in τ, and that ghk in τ implies that gk is in τ or hk is in τ.

11.6. Corollary. Let θ be a homomorphism from a commutative Banach algebra A into a Banach algebra B such that $\mathfrak{S}(\theta) = B$. Then there is a closed ideal K in B such that $K \cap \theta(A)$ is the set of nilpotent elements in $\theta(A)$.

Proof. Let K be the intersection of all the closed ideals J corresponding to non-nilpotents $\theta(f)$ in B given by 11. 4.

11. 7. **Theorem.** Let Ψ be a locally compact Hausdorff space, and let θ be a discontinuous homomorphism from $C_0(\Psi)$ onto a dense subalgebra of a radical Banach algebra B. Then

(i) $(\theta(f)B)^- = (\theta(f)^n B)^-$ for all positive integers n, and $\theta(f)$ is non-nilpotent for all f in $C_0(\Psi)$,

(ii) $\{(\theta(f)B)^- : f \in C_0(\Psi)\}$ satisfies the descending chain condition, and

(iii) for each h in $C_0(\Psi)$ with $\theta(h) \neq 0$, there is a k in $C_0(\Psi)$ such that $Q\theta$ is a discontinuous homomorphism whose kernel is a prime ideal in $C_0(\Psi)$ and $Q\theta(h) \neq 0$, where $Q : B \to B/J : b \mapsto b + J$ and $J = \{b \in B : b\theta(k)B = \{0\}\}$.

Proof. Let g be a function with compact support in Ψ, and let m be in $C_0(\Psi)$ with $mg = g$. If $\theta(g)$ is non-zero, then $(\theta(m) - 1)\theta(g) = 0$ implies that 1 is in the spectrum of $\theta(m)$. This contradicts the hypothesis that B is a radical algebra. Thus θ annihilates functions with compact support, and as these are dense in $C_0(\Psi)$ we have $\mathfrak{S}(\theta) = B$. We shall use this equality frequently.

(i) We apply Lemma 11. 2 with $X = C_0(\Psi)$, $Y = B$, $\mathcal{G}$ the semigroup of real valued non-negative functions in $C_0(\Psi)$, $T(g)$ multiplication by g, and $R(g)$ multiplication by $\theta(g)$ for all g in $\mathcal{G}$. Then $\{(\theta(f)B)^- : f \in \mathcal{G}\}$ satisfies the descending chain condition, and $(\theta(f)^n B)^- = (\theta(f)B)^-$ for all f in $\mathcal{G}$ and all positive integers n. We now prove that $\mathcal{G}$ may be replaced by $C_0(\Psi)$. If f is in $C_0(\Psi)$, then $|f|$ is in $\mathcal{G}$ and

$$(\theta(f)\mathfrak{S})^- \supseteq (\theta(f\bar{f})\mathfrak{S})^- = (\theta(|f|)^2\mathfrak{S})^-$$
$$= (\theta(|f|^{\frac{1}{2}})\mathfrak{S})^- \supseteq (\theta(|f|^{\frac{1}{2}}b)\mathfrak{S})^- = (\theta(f)\mathfrak{S})^-,$$

where

$$b(\lambda) = \begin{cases} 0 & \text{if } f(\lambda) = 0 \\ f(\lambda)/|f(\lambda)|^{\frac{1}{2}} & \text{if } f(\lambda) \neq 0 . \end{cases}$$

Note that b is in $C_0(\Psi)$. Thus $(\theta(f)B)^- = (\theta(|f|)B)^-$ for all f in $C_0(\Psi)$.

Hence $\{(\theta(f)B)^- : f \in \mathbf{C}_0(\Psi)\}$ satisfies the descending chain condition, and $(\theta(f)B)^- = (\theta(f)^n B)^-$ for all positive integers n and all f in $\mathbf{C}_0(\Psi)$. If $\theta(f)^n = 0$, then $\{0\} = (\theta(f)B)^- = (\theta(|f|^{1/4})B)^-$ so $\theta(|f|^{\frac{1}{2}}) = 0$. This implies that $\theta(f) = \theta(|f|^{\frac{1}{2}}b) = 0$. This proves (i) and (ii).

Property (iii) follows from Theorem 11.4 since $\theta(h)$ is not nilpotent.

11.8. Remarks. (a) In Theorem 10.4(ii) we showed that $\mu = \theta$ on $\tau^- . \tau$. This can be strengthened to $\mu = \theta$ on τ. If f is in τ, then $\theta(f)\mathfrak{S} = (\mu(f) + \nu(f))\mathfrak{S} = \nu(f)\mathfrak{S} = \{0\}$. Applying 11.7(i) to $\mathbf{C}_0(\Omega\backslash F)$, ν, and f it follows that $\nu(f) = 0$ so that $\theta(f) = \mu(f)$.

(b) The results in this section are taken from [118].

12. Homomorphisms and derivations from C*-algebras

12.1. Lemma. _Let_ J _be a closed two sided ideal in a_ C*-_algebra_ A. _If_ $J \cap C^*(a)$ _is of finite codimension in_ $C^*(a)$ _for each_ C*-_subalgebra_ $C^*(a)$ _of_ A _generated by a single hermitian element_ a _in_ A, _then_ J _is of finite codimension in_ A.

Proof. We may assume that A has an identity. Let $B = A/J$. Let D be a maximal commutative *-subalgebra of B. Then D is isometrically isomorphic to $\mathbf{C}(\Omega)$ for some compact Hausdorff space Ω. If x is hermitian in D, then $x = y + J$, where y is hermitian in A. By hypothesis $C^*(y) \cap J$ is of finite codimension in $C^*(y)$, so that there is a monic polynomial p such that p(y) is in J. Therefore $p(x) = 0$, and the spectrum of x is finite. Hence every element in $\mathbf{C}(\Omega)$ is finite valued so that Ω is finite. (If Ω were not finite we could construct a function with infinite range from the regularity of Ω.) Thus $D = \mathbf{C}e_1 \oplus \dots \oplus \mathbf{C}e_n$ where $e_i e_j = 0$ $(i \neq j)$, and $e_i^2 = e_i^* = e_i$ for all i. If $b = b^*$ is in B, then $(e_i b e_i)e_j = e_j(e_i b e_i)$ for all i and j, and $(e_i b e_i)^* = e_i b e_i$ so the maximality of D implies that $e_i b e_i$ is in D. Hence $e_i B e_i = \mathbf{C}e_i$, and e_i is a minimal idempotent for all i. The linear subspaces $e_i B e_j$ span B, and each has dimension one. Thus B is finite dimensional, and the proof is complete.

The above lemma is due to Ogawasara [97] and Johnson [64], and the following theorem is in [64].

12.2. Theorem. Let θ be a homomorphism from a C*-algebra A onto a dense subalgebra of a Banach algebra B. If

$$\tau = \{a \in A : \theta(a)\mathfrak{S} = \mathfrak{S}\theta(a) = \{0\}\},$$

then τ is a two sided ideal in A, and τ^- is of finite codimension in A.

Proof. Since $\mathfrak{S}$ is an ideal in B and θ is a homomorphism, τ is an ideal in A. Thus τ^- is an ideal in A. Let $a = a^*$ be in A, and let $C^*(a)$ be the commutative C*-algebra generated by a. Regard A as a Banach $C^*(a)$-module by left multiplication by elements of $C^*(a)$, and B as a $C^*(a)$-module by $f.b = \theta(f)b$ for all f in $C^*(a)$ and b in B. Then there is a finite subset F_1 of the carrier space of $C^*(a)$ such that $J(F_1)\mathfrak{S} = \{0\}$ by Theorem 9.3. Repeating this with right multiplication in place of left we obtain a finite subset F_2 of the carrier space of $C^*(a)$ such that $\mathfrak{S}.J(F_2) = \{0\}$. Thus $J(F_1 \cup F_2)$ is contained in τ, so that $\tau^- \cap C^*(a)$ has finite codimension in $C^*(a)$. By Lemma 12.1 the closure of τ is of finite codimension in A. This completes the proof.

12.3. Remark. We use the notation of Theorem 12.2. The ideal τ is closed if and only if θ is continuous. If θ is continuous, then clearly τ is closed. Conversely suppose that τ is closed. Let $x_n \to 0$ in τ. Then there are $y_1, y_2, \ldots$ and a in τ such that $x_n = ay_n$ for all n, and $y_n \to 0$ as $n \to \infty$ (either by the Varopoulos-Johnson version of Cohen's Factorization Theorem [49] or more simply by Ringrose [105, p. 434]). Since $\theta(a)\mathfrak{S} = \{0\}$, it follows that $x \mapsto \theta(a)\theta(x)$ is continuous, and hence $\theta(x_n) = \theta(a)\theta(y_n) \to 0$ as $n \to \infty$. Therefore θ is continuous on τ. Because τ is closed and of finite codimension in A, the homomorphism θ is continuous on A.

12.4. Corollary. If a unital C*-algebra A has no proper closed ideal of finite codimension, then every homomorphism from A into a Banach algebra is continuous.

Proof. Since A has no proper closed ideals of finite codimension,

$\tau^- = A$. Thus $\tau = A$, and θ is continuous by 12.3, or because $\theta(1)\mathfrak{S} = 1.\ \mathfrak{S} = \mathfrak{S} = \{0\}$.

The algebra of all bounded linear operators on an infinite dimensional Hilbert space satisfies the hypotheses of Corollary 12.4.

12.5. Corollary. *Let* A *be a* C^**-algebra, and let* X *be a Banach* A*-bimodule. If* δ *is a derivation from* A *into* X, *then* δ *is continuous.*

Proof. We apply the construction of Remark 8.1 to A, X, and δ. Then $\theta : A \to A \oplus X$ defined by $\theta(a) = (a, \delta(a))$ is a homomorphism, and $\mathfrak{S}(\theta) = \{0\} \oplus \mathfrak{S}(\delta)$. Hence

$$\begin{aligned}\tau &= \{a \in A : \theta(a)\mathfrak{S}(\theta) = \mathfrak{S}(\theta)\theta(a) = \{0\}\} \\ &= \{a \in A : a\mathfrak{S}(\delta) = \mathfrak{S}(\delta)a = \{0\}\},\end{aligned}$$

so τ is closed in A because X is a Banach A-bimodule. Thus θ and δ are continuous by 12.3.

12.6. Lemma. *Let* τ *be a two sided ideal in a* C^**-algebra* A *that intersects each* C^**-subalgebra* $C^*(a)$ *of* A *generated by a single hermitian element* a *in* A *in a closed ideal of finite codimension in* $C^*(a)$. *Then* τ *is closed and of finite codimension in* A.

Proof. Clearly τ^- is a closed ideal in A such that $\tau^- \cap C^*(a)$ is of finite codimension in $C^*(a)$ for each hermitian a in A. Hence τ^- is of finite codimension in A (Lemma 12.1). We now prove that $\tau = \tau^-$. Since a closed ideal in a C^*-algebra is a $*$-ideal [34], $\tau^- = \{x + iy : x = x^*,\ y = y^*;\ x, y \in \tau\}$. Let $x = x^*$ be in τ^-. Since $C^*(x) \cap \tau$ is a closed ideal of finite codimension in $C^*(x)$ there is a polynomial p with no repeated factors such that $p(x)$ is in τ. Further τ^-/τ is a radical algebra since an irreducible representation of τ^-/τ would give rise to a primitive, and hence closed, ideal in τ^- containing τ so contradicting the density of τ in τ^-. The linear factors $\lambda 1 - x$ with $\lambda \neq 0$ may be cancelled from p modulo τ. Hence x is in τ or 1 is in τ. Thus $\tau = \tau^-$, and the proof is complete.

12.7. Theorem. Let θ be a homomorphism from a C*-algebra A into a Banach algebra B. If θ is continuous on C*-subalgebras of A generated by single hermitian elements, then θ is continuous on A.

Proof. Let τ be defined as in Theorem 12.2. For each hermitian element a in A, $(\tau \cap C^*(a))^-$ is of finite codimension in $C^*(a)$ as we proved in 12.2 using 9.3. Because θ is continuous on $C^*(a)$, it follows that $\tau \cap C^*(a)$ is closed. Hence τ is closed by Lemma 12.6, and so θ is continuous by 12.3.

12.8. Corollary. If there is a discontinuous homomorphism from a C*-algebra into a Banach algebra, then there is a discontinuous homomorphism from $\mathbf{C}[0, 1]$ into a Banach algebra.

Proof. Suppose that every homomorphism from $\mathbf{C}[0, 1]$ into a Banach algebra is continuous. Let θ be a homomorphism from a C*-algebra A into a Banach algebra. If x is a hermitian element in A, then there is a continuous open homomorphism from $\mathbf{C}[0, 1]$ onto $C^*(a)$. The product of this homomorphism and the restriction of θ to $C^*(a)$ is continuous. Hence the restriction of θ to $C^*(a)$ is continuous. By Theorem 12.7, θ is continuous. This proves Corollary 12.8.

12.9. Remarks. (a) The algebra $\mathbf{C}[0, 1]$ in 12.8 may be replaced by $\mathbf{C}(\Omega)$, where Ω is a compact Hausdorff space containing a subset homeomorphic to the one point compactification of the integers $\mathbf{Z}$. We outline the proof. For such Ω there is an open continuous homomorphism from $\mathbf{C}(\Omega)$ onto $c_0 \oplus \mathbf{C}1$. It is thus sufficient to prove 12.8 with $\mathbf{C}[0, 1]$ replaced by $c_0 \oplus \mathbf{C}1$. If there is a discontinuous homomorphism from a C*-algebra, then there is a discontinuous homomorphism χ from $\mathbf{C}_0(\mathbf{R})$ that annihilates functions with compact support (12.8 and 10.4). There are two closed ideals K and L in $\mathbf{C}_0(\mathbf{R})$ each derived from a countable family of disjoint open intervals such that $K + L = \mathbf{C}_0(\mathbf{R})$, and K, L, $\theta(K)$, $\theta(L)$ may be regarded as Banach $c_0 \oplus \mathbf{C}1$-modules. Corollary 9.4 and some technical working shows that χ is continuous on K and L. This completes the outline of the proof.

(b) The following references contain the results of this section and related results: Bade and Curtis [10], Cleveland [24], Dixmier [34], [35], Johnson [60], [64], [66], Laursen and Stein [78], Ringrose [105], Sakai [108, [109], Sinclair [117], Stein [122], [123], Laursen and Sinclair [148].

Linear operators

Introduction

The single section in this chapter contains some results on the automatic continuity of positive linear functionals on Banach *-algebras. The classical result on the automatic continuity of positive linear functionals is that a positive linear functional on a unital Banach *-algebra with continuous involution is continuous (Corollary 13.3). A problem to replace the hypothesis that there is an identity by a weaker assumption still implying the automatic continuity of positive linear functionals. An initial conjecture was that the linear subspace A^2 of A spanned by $a_1 a_2$ for all a_1, a_2 in A must be closed and of finite codimension in A (Varopoulos [126]). Certainly when A is commutative and A^3 is closed and of finite codimension in A, then every positive linear functional on A is continuous (Varopoulos [126], Murphy [92], Theorem 13.7). In Theorem 13.9 we prove a result due to J. Cusack (see also Varopoulos [126], Ando [143]): if A^2 is closed and of finite codimension in A and if

$$A^+ = \{\sum_1^n a_j^* a_j : a_1, \ldots, a_n \in A \text{ for all } n\}$$

is closed, then every positive linear functional on A is continuous. The section ends with Varopoulos's proof that a positive linear functional on a Banach *-algebra with bounded approximate identity is continuous.

13. Positive linear functionals

Throughout this section let A be a Banach *-algebra - a Banach algebra with a conjugate linear involution * on the algebra. A linear functional f on a Banach *-algebra A is said to be <u>positive</u> if $f(a^*a) \geq 0$ for all a in A. We let A^n denote the linear space spanned by $x_1 \ldots x_n$ for all $x_1, \ldots, x_n$ in A, and let f be a positive linear functional on A.

The problem of automatic continuity of positive linear functionals is to find necessary and sufficient conditions on a Banach *-algebra A so that every positive linear functional on A is continuous. The first result on the automatic continuity of homomorphisms was that multiplicative linear functionals on a unital Banach algebra are continuous. The starting point for positive linear functionals is the theorem that every positive linear functional on a unital Banach algebra with continuous involution is continuous.

The first lemma is Ford's square root lemma [44], [18].

13.1. Lemma. *Let $a = a^*$ be in A with $\sigma(a) \cap [1, \infty) = \emptyset$. Then there is a unique $x = x^*$ in A satisfying $\sigma(x) \subseteq \{z \in \mathbf{C} : \mathrm{Re}\, z < 1\}$ and $2x - x^2 = a$.*

Proof. Let A_1 be the Banach *-algebra obtained from A by adjoining an identity, and let $f(z) = 1 - (1 - z)^{\frac{1}{2}}$ be analytic in the domain $\mathbf{C} \setminus [1, \infty)$. Let $x = f(a)$ be defined by the analytic functional calculus. Then $(1 - x)^2 = 1 - a$ and $\sigma(x) \subseteq \{z \in \mathbf{C} : \mathrm{Re}\, z < 1\}$. Hence x^* also satisfies $(1 - x^*)^2 = 1 - a$ and $\sigma(x^*) \subseteq \{z \in \mathbf{C} : \mathrm{Re}\, z < 1\}$. Let y be in A with $\sigma(y) \subseteq \{z \in \mathbf{C} : \mathrm{Re}\, z < 1\}$ and $(1 - y)^2 = 1 - a$. We shall show that $y = x$ which will complete the proof. Since $ya = ay$ we have $yx = xy$ from the definition of x. Hence $\sigma(x + y) \subseteq \{z \in \mathbf{C} : \mathrm{Re}\, z < 2\}$, and so $x + y - 2$ is invertible in A. Also $(1 - x)^2 = 1 - a = (1 - y)^2$ implies that $0 = (x - y)(x + y - 2)$ so that $x = y$, and the proof is complete.

Let $\nu(x)$ denote the spectral radius of an element x in a Banach algebra.

13.2. Lemma. *If a, b, $x = x^*$ are in A, and if f is a positive linear functional on A, then*

(i) $f(a^*b) = f(b^*a)^-$,

(ii) $|f(a^*b)|^2 \leq f(a^*a)f(b^*b)$,

(iii) $|f(a^*xa)| \leq f(a^*a)\nu(x)$, *and*

(iv) $|f(a^*ba)| \leq f(a^*a)\nu(b^*b)^{\frac{1}{2}}$.

Proof. Let α and β be in $\mathbf{C}$, and apply f to $(\alpha a+\beta b)^*(\alpha a+\beta b)$.

Then

$$|\alpha|^2 f(a^*a) + \bar{\alpha}\beta f(a^*b) + \alpha\bar{\beta} f(b^*a) + |\beta|^2 f(b^*b) \geq 0.$$

Suitable choices of α and β give (i) and (ii), in the same way as these inequalities are obtained for an inner product.

Assume that $\nu(x) < 1$. By Lemma 13.1 there are y and z in A with $y = y^*$, $z = z^*$, $x = 2y - y^2$, and $-x = 2z - z^2$. Let $v = a - ya$, and $w = a - za$. Then

$$v^*v = a^*(1 - y)^2 a = a^*(1 - x)a,$$

and

$$w^*w = a^*(1 - z)^2 a = a^*(1 + x)a,$$

so that

$$f(a^*a) - f(a^*xa) = f(v^*v) \geq 0,$$

and

$$f(a^*a) + f(a^*xa) = f(w^*w) \geq 0.$$

Therefore $|f(a^*xa)| \leq f(a^*a)$, and we have proved (iii). By (ii), $|f(a^*ba)|^2 = |f(a^*(ba))|^2 \leq f(a^*a)f(a^*b^*ba)$ so that $|f(a^*ba)|^2 \leq f(a^*a)^2 \nu(b^*b)$ by (iii). This completes the proof.

13.3. **Corollary.** _There is a constant_ M _such that_ $|g(a^*ba)| \leq Mg(a^*a)\|b\|$ _for all_ a _and_ b _in_ A, _and all positive linear functionals_ g _on_ A. _If_ A _has an identity, then every positive linear functional on_ A _is continuous._

Proof. Let R denote the radical of A. Then * induces an involution on A/R because $R = R^*$ which follows from the result that the left radical is equal to the right radical (Rickart [103, p. 55]). This involution on A/R is continuous because A/R is semisimple (6.13). Hence there is a constant M^2 so that $\|x^* + R\| \leq M^2\|x + R\|$ for all x in A. Therefore

$$\nu(b^*b) = \nu(b^*b + R) \leq \|b^*b + R\| \leq M^2\|b + R\|^2 \leq M^2\|b\|^2$$

so that $|g(a^*ba)| \leq Mg(a^*a)\|b\|$ by 13.2(iv). If A has an identity, then

$|g(b)| \leq Mg(1)\|b\|$ for all b in A.

13.4. Example. If A^2 is not closed and of finite codimension in A, then there are discontinuous positive linear functionals on A. For, by Zorn's Lemma there are discontinuous linear functionals on A annihilating A^2, and any such linear functional is positive. A natural question arises from this observation and the result on images of Banach spaces under continuous linear operators (Lemma 3.3): if B is a Banach algebra and if the linear space B^2 is of finite codimension in B, is this linear space closed? This is related to the question on Exercise 11, Chapter 2 in Rudin [107, p. 375].

13.5. Remarks. A linear functional g on A is said to <u>dominate</u> a linear functional h on A if $g - h$ is positive, and we write $g \geq h$ if $g - h$ is positive. Using the identity

$$4axb = (b+a^*)^*x(b+a^*) - (b-a^*)^*x(b-a^*) + i(b+ia^*)^*x(b+ia^*) - i(b-ia^*)^*x(b-ia^*)$$

with $x = 1$, an element in A^2 may be written in the form $\sum \alpha_j a_j^* a_j$ for some finite set $\{a_1, \ldots, a_n\}$ of elements of A, and complex numbers $\{\alpha_1, \ldots, \alpha_n\}$.

13.6. Lemma. *If A^2 is closed and if there is a discontinuous positive linear functional f on A^2, then there is a discontinuous positive linear functional g on A^2 that does not dominate any non-zero continuous positive linear functional and such that g is dominated by f.*

Proof. Let $\mathbf{G}$ be the set of non-zero continuous positive linear functionals on A^2 that are dominated by f. If $\mathbf{G}$ is empty, then we take g to be f so we may assume that $\mathbf{G}$ is non-empty. Now $\geq$ is a partial order on $\mathbf{G}$. Let $\mathbf{E}$ be a chain in $\mathbf{G}$. Then $\lim\{k(a^*a) : k \in \mathbf{E}\}$ exists for each a in A because $k(a^*a) \leq f(a^*a)$ for all k in $\mathbf{E}$. We let $h(x) = \sum \alpha_j \lim k(a_j^* a_j)$ for each $x = \sum \alpha_j a_j^* a_j$ in A^2. Then h is a well defined positive linear functional on A^2 dominating the members of $\mathbf{E}$. For each x in A^2 the set $\{|k(x)| : k \in \mathbf{E}\}$ is bounded because

$$|k(x)| \le \sum_1^n |\alpha_j| k(a_j^* a_j) \le \sum_1^n |\alpha_j| f(a_j^* a_j)$$

where $x = \sum_1^n \alpha_j a_j^* a_j$. The uniform boundedness theorem implies that h is continuous. Hence G satisfies the conditions of Zorn's Lemma. Let m be a maximal element of G, and let $g = f - m$ on A^2. Then g satisfies the conclusions of the lemma.

13.7. **Theorem.** *Let A be a Banach *-algebra with centre Z. If the linear space Z^2A spanned by all elements $z_1 z_2 a$, with z_1, z_2 in Z and a in A, is closed and of finite codimension in A, then each positive linear functional on A is continuous.*

Proof. Because Z^2A is closed and of finite codimension in A it is sufficient to prove that a positive linear functional on A is continuous on Z^2A. By Lemma 13.6 it is sufficient to prove that a non-zero positive linear functional on A^2 dominates a continuous non-zero positive linear functional on A^2. Let f be a positive linear functional on A^2 that is non-zero on Z^2A. Then there is an a in Z and an x in A such that $f(a^*xa) \neq 0$ by Remark 13.5. We may suppose that $\|a^*a\| < 1$. Then there is $b = b^*$ in A such that $(1 - b)^2 = 1 - a^*a$ by Lemma 13.1. If $f_a(c) = f(a^*ca)$ for all c in A, then

$$\begin{aligned}(f - f_a)(y^*y) &= f(y^*y - a^*y^*ya)\\ &= f(y^*(1 - a^*a)y)\\ &= f(y^*(1 - b^*)(1-b)y) \ge 0\end{aligned}$$

for all y in A. Thus f dominates f_a, and f_a is non-zero and continuous by Corollary 13.3. This proves Theorem 12.7.

Let $A^+ = \{\sum a_j^* a_j : \{a_1, \ldots, a_n\}$ is a finite subset of $A\}$.

13.8. **Lemma.** *If A^+ is closed and if f is a positive linear functional on A, then there is a constant M such that $f(x) \le M\|x\|$ for all x in A^+.*

Proof. Suppose that there is no such constant M. Choose a sequence $\{x_n\}$ in A^+ such that $f(x_n) > 2^n\|x_n\|$ for all n, and let

$y_m = \sum_{n=m}^{\infty} 2^{-n}\|x_n\|^{-1}x_n$ for $m = 1, 2, \ldots$. Then y_m is in A^+ for all m, and $y_1 = \sum_{n=1}^{m-1} 2^{-n}\|x_n\|^{-1}x_n + y_m$ for all $m \geq 2$. Hence $f(y_1) \geq f(\sum_{n=1}^{m-1} 2^{-n}\|x_n\|^{-1}x_n) \geq m - 1$ for all m. This contradiction proves the lemma.

13. 9. Theorem. _Let_ A _be a Banach_ *-_algebra, and let_ A^+ _be the set of all finite sums of elements_ a*a _with_ a _in_ A. _If_ A^2 _is closed and of finite codimension in_ A, _and if_ A^+ _is closed, then each positive linear functional on_ A _is continuous._

Proof. We shall prove that there is an N such that each a in A^2 may be written $a = z_1 - z_2 + i(z_3 - z_4)$ where $\|z_j\| \leq N\|a\|$ and z_j is in A^+ for $j = 1, \ldots, 4$. Theorem 13. 9 will follow from this by using Lemma 13. 8. The following proof is an adaptation of the proof of the open mapping theorem. Let $Y = \{(x_1, \ldots, x_4) : x_j \in A^+\}$, let $Y_\alpha = \{(x_1, \ldots, x_4) \in Y : \|x_j\| \leq \alpha\}$, let $A^2_\alpha = \{a \in A^2 : \|a\| \leq \alpha\}$, and let $T : Y \to A^2$ be defined by

$$T(x_1, \ldots, x_4) = x_1 - x_2 + i(x_3 - x_4).$$

By 13. 5, TY is equal to A^2 and so $A^2 = \bigcup_{n=1}^{\infty} (TY_n)^-$. The Baire Category Theorem implies that some $(TY_n)^-$ has non empty interior, and hence 0 is in the interior of $(TY_{2n})^-$. Therefore there is a $\beta > 0$ such that $(TY_{\alpha\beta})^- \supseteq A^2_\alpha$ for all $\alpha > 0$. Let a be in A^2 with $\|a\| \leq 1$. By induction we choose a sequence $\{y_n\}$ in Y such that

(i) $\|T(y_1 + \ldots + y_n) - a\| < 2^{-n}$, and

(ii) $y_n \in Y_{\beta. 2^{-n+1}}$.

If $y_n = (x_{1n}, \ldots, x_{4n})$, we let $z_j = \sum_{n=1}^{\infty} x_{jn}$ for $j = 1, \ldots, 4$. Then $\|z_j\| \leq 2\beta$ for each j, and $a = z_1 - z_2 + i(z_3 - z_4)$. This completes the proof.

We shall require the following version of Cohen's Factorization Theorem in the proof of Theorem 13. 11. Proofs of this result may be found in [104], [18], [49].

13.10. Lemma. *Let A be a Banach algebra with a bounded approximate identity. If (x_n) is a sequence in A with $x_n \to 0$ as $n \to \infty$, then there are $a, y_1, y_2, \ldots$ in A such that $x_n = ay_n$ for all n and $y_n \to 0$ as $n \to \infty$.*

13.11. Theorem. *Let A be a Banach *-algebra. If A has a bounded approximate identity, then each positive linear functional on A is continuous.*

Proof. Let f be a positive linear functional on A, and let (x_n) be a sequence in A with $x_n \to 0$ as $n \to \infty$. Then there are $a, y_1, y_2, \ldots$ in A such that $x_n = ay_n$ for all n and $y_n \to 0$ as $n \to \infty$ (Lemma 13.10). By the right multiplication version of Lemma 13.10 there are $b, z_1, z_2, \ldots$ in A such that $y_n = z_n b$ for all n and $z_n \to 0$ as $n \to \infty$. If $F(x) = f(axb)$ for all x in A, then F is continuous by 13.3 and 13.5. Hence $f(x_n) = F(z_n) \to 0$ as $n \to \infty$, and the proof is complete.

Bibliography

[1] G. R. Allan. Embedding the algebra of formal power series in a Banach algebra, Proc. London Math. Soc. (3) 25 (1972) 329-40.

[2] G. R. Allan. Elements of finite closed descent in a Banach algebra, J. London Math. Soc. (2) 7 (1973) 462-4.

[3] M. F. Atiyah and I. G. Macdonald. Introduction to commutative algebra, Addison-Wesley (1969).

[4] G. F. Bachelis. Homomorphisms of annihilator Banach algebras, Pacific J. Math. 25 (1968) 229-47.

[5] G. F. Bachelis. Homomorphisms of annihilator Banach algebras II, Pacific J. Math. 30 (1969) 283-91.

[6] G. F. Bachelis. Homomorphisms of Banach algebras with minimal ideals, Pacific J. Math. 41 (1972) 307-12.

[7] W. G. Bade and P. C. Curtis, Jr. Homomorphisms of commutative Banach algebras, American J. Math. 82 (1960) 589-608.

[8] W. G. Bade and P. C. Curtis, Jr. The Wedderburn decomposition of commutative Banach algebras, American J. Math. 82 (1960) 851-66.

[9] W. G. Bade and P. C. Curtis, Jr. Embedding theorems for commutative Banach algebras, Pacific J. Math. 18 (1966) 391-409.

[10] W. G. Bade and P. C. Curtis, Jr. Continuity of derivations of Banach algebras, J. Functional Analysis, to appear.

[11] B. Bainerd and R. E. Edwards. Linear operators which commute with translations I, II, J. Australian Math. Soc. 6 (1966) 289-350.

[12] B. A. Barnes. Some theorems concerning the continuity of algebra homomorphisms, Proc. Amer. Math. Soc. 18 (1967) 1035-7.

[13] B. A. Barnes. Linear functional continuous on abelian *-subalgebras of a B*-algebra, to appear.

[14] B. A. Barnes and J. Duncan. The Banach algebra $l^1(S)$, to appear.

[15] T. K. Boehme. Continuity and perfect operators, J. London Math. Soc. 39 (1964) 355-8.

[16] F. F. Bonsall. A minimal property of the norm in some Banach algebras, J. London Math. Soc. 29 (1954) 156-64.

[17] F. F. Bonsall. A survey of Banach algebra theory, Bull. London Math. Soc. 2 (1970) 257-74.

[18] F. F. Bonsall and J. Duncan. Complete normed algebras, Springer-Verlag (1973).

[19] N. Bourbaki. Éléments de mathematique. Topologie générale, Chapter I-II, Hermann (1961).

[20] D. T. Brown. A class of Banach algebras with unique norm topology, Proc. Amer. Math. Soc. 17 (1966) 1429-34.

[21] R. L. Carpenter. Continuity of derivations in F-algebras, Amer. J. Math. (197) 500-2.

[22] H. Cartan and S. Eilenberg. Homological algebra, Princeton Univ. Press (1956).

[23] P. Civin and B. Yood. Lie and Jordan structures in Banach algebras, Pacific J. Math. 15 (1965) 775-97.

[24] S. B. Cleveland. Homomorphisms of non-commutative *-algebras, Pacific J. Math. 13 (1963) 1097-1109.

[25] P. J. Cohen. Factorization in group algebras, Duke Math. J. 26 (1959) 199-205.

[26] I. Colojoară and C. Foias. Theory of generalized spectral operators, Gordon and Breach (1968).

[27] P. C. Curtis. Derivations of commutative Banach algebras, Bull. Amer. Math. Soc. 67 (1961) 271-3.

[28] H. G. Dales. The uniqueness of the functional calculus, Proc. London Math. Soc. (3) 27 (1973) 638-48.

[29] H. G. Dales. Exponentiation in Banach star algebras, Preprint.

[30] H. G. Dales and J. P. McClure. Continuity of homomorphisms into certain commutative Banach algebras, Proc. London Math. Soc. (3) 26 (1973) 69-81.

[31] H. G. Dales and J. P. McClure. Completion of normed algebras of polynomials, to appear.

[32] H. G. Diamond. Characterization of derivations on an algebra of measures. Math. Z. 100 (1967) 135-40.

[33] H. G. Diamond. Characterization of derivations on an algebra of measures II, Math. Z. 105 (1968) 301-6.

[34] J. Dixmier. Les C*-algèbres et leurs representations, Gauthier-Villars (1968).

[35] J. Dixmier. Les algèbres d'opérateurs dans l'espace Hilbertien, Gauthier-Villars (1969).

[36] P. G. Dixon and D. H. Fremlin. A remark concerning multiplicative functionals on LMC algebras, J. London Math. Soc. (2) 5 (1972) 231-2.

[37] R. S. Donan. Does there exist more than one Banach *-algebra with discontinuous involution, Amer. Math. Monthly 79 (1972) 762-4.

[38] R. M. Dudley. Continuity of homomorphs, Duke Math. J. 28 (1961) 587-93.

[39] J. Dugundji. Topology, Allyn and Bacon (1966).

[40] J. Duncan. The continuity of the involution on Banach *-algebras, J. London Math. Soc. 41 (1966) 701-6.

[41] M. Eidelheit. On isomorphisms of rings of linear operators, Studia Math. 9 (1940) 97-105.

[42] J. Esterle. Normes d'algèbres minimales, topologie d'algèbre normée minimum sur certaines algèbres d'endomorphismes continus d'un espace norme, C. R. Acad. Sci. Paris, Ser. A, 277 (1973) 425-7.

[43] C. Feldman. The Wedderburn principal theorem in Banach algebras, Proc. American Math. Soc. 2 (1951) 771-7.

[44] J. W. M. Ford. A square root lemma for Banach *-algebras, J. London Math. Soc. 42 (1967) 521-2.

[45] I. M. Gelfand. Normierte Ringe, Math. Sbornik 9 (1941) 3-24.

[46] T. A. Gillespie and T. T. West. Operators generating weakly compact groups, Proc. Royal Irish Academy 73 (1973) 309-26.

[47] S. Grabiner. Ranges of products of operators, Canadian J. Math. to appear.

[48] P. Gvozdková. On continuity of linear transformations commuting with generalized scalar operators, Comm. Math. Univ. Carol. 11 (1970) 583-8.

[49] E. Hewitt and K. A. Ross. Abstract harmonic analysis, Vol. II, Springer (1970).

[50] N. Jacobson. Lectures in abstract algebra, Vol. III, Van Nostrand (1955).

[51] N. Jacobson. Structure of rings, Amer. Math. Soc. Colloqu. 37 Amer. Math. Soc. (1956).

[52] T. Jimbo, K. Tsurumi, and K. Izuchi. Continuity of H-derivations on commutative Banach algebras, Sci. Reports Tokyo Kyoiku Doigaku, Section A, 11 (1972) 6-9.

[53] B. E. Johnson. Continuity of homomorphisms of topological algebras, Proc. Cambridge Phil. Soc. 60 (1964) 171-2.

[54] B. E. Johnson. An introduction to the theory of centralizers, Proc. London Math. Soc. (3) 14 (1964) 299-320.

[55] B. E. Johnson. Centralizers on certain topological algebras, J. London Math. Soc. 39 (1964) 603-14.

[56] B. E. Johnson. Continuity of centralizers on Banach algebras, J. London Math. Soc. 41 (1966) 639-40.

[57] B. E. Johnson. Continuity of transformations which leave invariant certain translation invariant subspaces, Pacific J. Math. 20 (1967) 223-30.

[58] B. E. Johnson. Continuity of linear operators commuting with continuous linear operators, Trans. Amer. Math. Soc. 128 (1967) 88-102.

[59] B. E. Johnson. The uniqueness of the (complete) norm topology, Bull. Amer. Math. Soc. 73 (1967) 537-9.

[60] B. E. Johnson. Continuity of homomorphisms of algebras of operators, J. London Math. Soc. 42 (1967) 537-41.

[61] B. E. Johnson. Centralizers and operators reduced by maximal ideals, J. London Math. Soc. 43 (1968) 231-3.

[62] B. E. Johnson. The Wedderburn decomposition of Banach algebras with finite dimensional radical, Amer. J. Math. 90 (1968) 866-76.

[63] B. E. Johnson. Continuity of derivations on commutative algebras,

Amer. J. Math. 91 (1969) 1-10.

[64] B. E. Johnson. Continuity of homomorphisms of algebras of operators II, J. London Math. Soc. (2) 1 (1969) 81-4.

[65] B. E. Johnson. Cohomology in Banach algebras, Mem. American Math. Soc. 127 (1972).

[66] B. E. Johnson. Norming $C(\Omega)$ and related algebras, to appear.

[67] B. E. Johnson and S. K. Parrott. Operators commuting with a von Neumann algebra modulo the set of compact operators, J. Functional Analysis 11 (1972) 73-96.

[68] B. E. Johnson and A. M. Sinclair. Continuity of derivations and a problem of Kaplansky, American J. Math. 90 (1968) 1067-73.

[69] B. E. Johnson and A. M. Sinclair. Continuity of linear operators commuting with continuous linear operators II, Trans. American Math. Soc. 146 (1969) 533-40.

[70] R. V. Kadison. Derivation of operator algebras, Ann. Math. 83 (1966) 280-93.

[71] R. R. Kallman. The topology of compact simple Lie group is essentially unique, Advances in Math. 12 (1974) 416-17.

[72] I. Kaplansky. Normed algebras, Duke Math. J. 16 (1949) 399-418.

[73] I. Kaplansky. Modules over operator algebras, American J. Math. 75 (1953) 839-58.

[74] I. Kaplansky. Ring isomorphisms of Banach algebras, Canadian J. Math. 6 (1954) 374-81.

[75] I. Kaplansky. Derivations on Banach algebras. Seminars on analytic functions, Vol. 2, Institute for Advanced Study (1957).

[76] K. B. Laursen. On discontinuous homomorphisms of $L'(G)'$, Math. Scand. 30 (1972) 263-6.

[77] K. B. Laursen. A note on lifting of matrix units in C^*-algebras, Math. Scand. 33 (1973) 338-42.

[78] K. B. Laursen and J. D. Stein, Jr. Automatic continuity in Banach spaces and algebras, American J. Math. 95 (1974) 485-506.

[79] J. A. Lindberg. A class of commutative Banach algebras with unique complete norm topology and continuous derivations, Proc. American Math. Soc. 29 (1971) 516-20.

[80] R. J. Loy. Continuity of derivations on topological algebras of power series, Bull. Australian Math. Soc. 1 (1969) 419-24.

[81] R. J. Loy. Uniqueness of the complete norm topology and continuity of derivations on Banach algebras, Tôhoku Math. J. 22 (1970) 371-8.

[82] R. J. Loy. Continuity of higher derivations, Proc. American Math. Soc. 37 (1973) 505-10.

[83] R. J. Loy. Commutative Banach algebras with non-unique norm topology, Bull Australian Math. Soc., to appear.

[84] R. J. Loy. Continuity of linear operators commuting with shifts, to appear.

[85] R. J. Loy. Uniqueness of the Fréchet space topology on certain topological algebras, to appear.

[86] E. H. Luchins. On radicals and continuity of homomorphisms into Banach algebras, Pacific J. Math. 9 (1959) 755-8.

[87] J. P. McClure. Automatic continuity of functional calculus for vector valued functions, J. London Math. Soc. (2) 5 (1972) 154-8.

[88] G. H. Meisters. Translation invariant linear forms and a formula for the Dirac measure, J. Functional Analysis 8 (1971) 173-88.

[89] G. H. Meisters. Some discontinuous translation invariant linear forms, J. Functional Analysis 12 (1973) 199-210.

[90] G. H. Meisters and W. M. Schmidt. Translation invariant linear forms on $L^2(G)$ for compact abelian groups G, J. Functional Analysis 11 (1972) 407-24.

[91] J. Moffat. Continuity of automorphic representations, Proc. Cambridge Phil. Soc. 74 (1973) 461-5.

[92] I. S. Murphy. Continuity of positive linear functionals on Banach *-algebras, Bull. London Math. Soc. 1 (1969) 171-3.

[93] M. A. Naĭmark. Normed rings, Noordhoff (1959).

[94] I. Namioka. Partially ordered linear topological spaces, American Math. Soc. Memoir No. 24, American Math. Soc. (1957).

[95] J. D. Newburgh. The variation of spectra, Duke Math. J. 18 (1951) 165-76.

[96] Shu-Bun Ng and S. Warner. Continuity of positive and multiplicative functionals, Duke Math. J. 39 (1972) 281-4.

[97] T. Ogasawara. Finite dimensionality of certain Banach algebras, J. Sci. Hiroshima Univ. 17 (1954) 359-64.

[98] A. L. Peressini. Ordered topological vectors spaces, Harper and Row (1967).

[99] V. Pták. A uniform boundedness theorem and mappings into spaces of operators, Studia Math. 31 (1968) 425-31.

[100] V. Pták and P. Vrbová. On the spectral function of a normal operator, Czech. Math. J. 23 (1973) 615-6.

[101] C. E. Rickart. The uniqueness of norm problem in Banach algebras, Ann. Math. 51 (1950) 615-28.

[102] C. E. Rickart. Representation of certain Banach algebras on Hilbert space, Duke Math. J. 18 (1951) 27-39.

[103] C. E. Rickart. General theory of Banach algebras, Van Nostrand (1960).

[104] M. Rieffel. On the continuity of certain intertwining operators, centralizers, and positive linear functionals, Proc. American Math. Soc. 20 (1969) 455-7.

[105] J. R. Ringrose. Automatic continuity of derivations of operator algebras, J. London Math. Soc. (2) 5 (1972) 432-8.

[106] J. R. Ringrose. Linear functionals on operator algebras and their abelian subalgebras, J. London Math. Soc. (2) 7 (1974) 553-60.

[107] W. Rudin. Functional analysis, McGraw-Hill (1973).

[108] S. Sakai. On a conjecture of Kaplansky, Tohoku Math. J. 12 (1960) 31-3.

[109] S. Sakai. C*-algebras and W*-algebras, Springer (1971).

[110] H. N. Shapiro. On the convolution ring of arithmetic functions, Comm. Pure and App. Math. 25 (1972) 287-334.

[111] G. Šilov. On regular normed rings, Trudy Mat. Inst. im V. A. Steklova 21 Moskva, Akad. Nauk SSSR (1947).

[112] A. M. Sinclair. Continuous derivations on Banach algebras, Proc. American Math. Soc. 20 (1969) 166-70.

[113] A. M. Sinclair. Jordan homomorphisms on a semisimple Banach algebra, Proc. American Math. Soc. 24 (1970) 209-14.

[114] A. M. Sinclair. Jordan automorphisms on a semisimple Banach algebra, Proc. American Math. Soc. 25 (1970) 526-8.

[115] A. M. Sinclair. The Banach algebra generated by a hermitian operator, Proc. London Math. Soc. (3) 24 (1972) 681-91.

[116] A. M. Sinclair. A discontinuous intertwining operator, Trans. American Math. Soc. 188 (1974) 259-67.

[117] A. M. Sinclair. Homomorphisms from C*-algebras, Proc. London Math. Soc., to appear.

[118] A. M. Sinclair. Homomorphisms from $\mathbf{C}_0(\mathbf{R})$, submitted for publication.

[119] I. M. Singer and J. Wermer. Derivations on commutative normed algebras, Math. Ann. 129 (1955) 260-4.

[120] T. A. Slobko. Algebra norms on C(G), American J. Math. 92 (1970) 381-8.

[121] J. D. Stein, Jr. Homomorphisms of semisimple algebras, Pacific J. Math. 26 (1968) 589-94.

[122] J. D. Stein, Jr. Continuity of homomorphisms of von Neumann algebras, American J. Math. 91 (1969) 153-9.

[123] J. D. Stein, Jr. Homomorphisms of B*-algebras, Pacific J. Math. 28 (1969) 431-40.

[124] B. L. D. Thorp. Operators which commute with translations, J. London Math. Soc. 39 (1964) 359-69.

[125] W. Tiller. P-commutative Banach *-algebras, Trans. American Math. Soc. 180 (1973) 327-36.

[126] N. T. Varopoulos. Sur les formes positives d'une algèbre de Banach, C. R. Acad. Sci. Paris, Serie A, 258 (1964) 2465-7.

[127] N. T. Varopoulos. Continuité des formes linéaires positives sur une algèbre de Banach avec involution, C. R. Acad. Sci. Paris, Serie A, 258 (1964) 1121-4.

[128] P. Vrbová. On the continuity of linear transformations commuting with generalized scalar operators in Banach space, Časopis pro pěstovani matematiky 97 (1972) 142-50.

[129] P. Vrbová. Structure of maximal spectral spaces of generalized

scalar operators, Czech. Math. J. 23 (1973) 493-6.

[130] L. J. Wallen. On the continuity of a class of unitary representations, Michigan Math. J. 16 (1969) 153-6.

[131] J. D. M. Wright. All operators on a Hilbert space are bounded, Bull. Amer. Math. Soc. 79 (1973) 1247-50.

[132] B. Yood. Topological properties of homomorphisms between Banach algebras, American J. Math. 76 (1954) 155-67.

[133] B. Yood. Homomorphisms on normed algebras, Pacific J. Math. 8 (1958) 373-81.

[134] B. Yood. Continuity for linear maps on Banach algebras, Studia Math. 31 (1968) 263-6.

[135] N. J. Young. Separate continuity and multilinear operations, Proc. London Math. Soc. (3) 26 (1973) 289-319.

[136] R. L. Carpenter. Continuity of systems of derivations on F-algebras, Proc. American Math. Soc. 30 (1971) 141-6.

[137] R. E. Edwards. Endomorphisms of function spaces which leave stable all translation invariant manifolds, Pacific J. Math. 14 (1964) 31-8.

[138] I. Kaplansky. Infinite abelian groups, Univ. of Michigan Press (1954).

[139] I. Kaplansky. Algebraic and analytic aspects of operator algebras, American Math. Soc. (1970).

[140] V. Pták. Mappings into spaces of operators, Comment. Math. Univ. Carolinae 9 (1968) 161-4.

[141] J. D. Stein, Jr. Some aspects of automatic continuity, Pacific J. Math. 50 (1974) 187-204.

[142] B. E. Johnson. Continuity of linear operators commuting with quasi-nilpotent operators, Indiana Univ. Math. J. 20 (1971) 913-5.

[143] T. Andô. On fundamental properties of Banach space with a cone, Pacific J. Math. 12 (1962) 1163-9.

[144] B. Hartley and T. O. Hawkes. Rings, modules and linear algebra, Chapman and Hall (1970).

[145] F. Gulick. Systems of derivations, Trans. American Math. Soc. 149 (1970) 465-88.

[146] S. Shivali. Representability of positive functionals, J. London

Math. Soc. (2) 3 (1971) 145-70.

[147] F. F. Bonsall and J. Duncan. Numerical ranges of operators on normed spaces and elements of normed algebras, London Math. Soc. Lecture Note Series 2 (1971) Cambridge Univ. Press.

[148] K. B. Laursen and A. M. Sinclair. Lifting matrix units in C*-algebras II, Math. Scand. (to appear).

[149] A. M. Sinclair. Corrigendum: Homomorphisms from C*-algebras (see [117]), Proc. London Math. Soc.

[150] N. P. Jewell and A. M. Sinclair. Epimorphisms and derivations on $L^1[0, 1]$ are continuous, Bull. London Math. Soc. (to appear).

[151] A. L. T. Paterson. A note on the continuity of positive functionals on a Banach *-algebra. Preprint.

[152] K. B. Laursen. Some remarks on automatic continuity. Preprint.

Index